Building Skills in Geography

Teacher's Annotated Edition

Richard G. Boehm, Ph.D.
Professor of Geography
Department of Geography and Planning
Southwest Texas State University
San Marcos, Texas

Larry L. Bybee
Former Secondary Social Studies Supervisor
Northside Independent School District
San Antonio, Texas

James F. Petersen, Ph.D.
Professor
Department of Geography and Planning
Southwest Texas State University
San Marcos, Texas

GLENCOE

McGraw-Hill

New York, New York   Columbus, Ohio   Mission Hills, California   Peoria, Illinois

Contents

Send all inquiries to:
GLENCOE/McGraw-Hill
936 Eastwind Drive
Westerville, OH 43081-3374

Printed in the United States of America

ISBN 0–02–823724–2 (Student Text-Workbook)
ISBN 0–02–823725–0 (Teacher's Annotated Edition)

4 5 6 7 8 9 10   024   02 01 00 99 98 97

About *Building Skills in Geography*

The Teacher's Annotated Edition of *Building Skills in Geography* is designed to assist you in using the student text in your classroom. The Teacher's Annotated Edition is divided into two parts.

The first part of the Teacher's Annotated Edition is this 16-page Teacher's Manual. This manual provides you with general information about the student text and suggests some teaching strategies for meeting varying ability levels. Also included in the Teacher's Manual are eleven reproducible outline maps with suggested activities.

The second part of the Teacher's Annotated Edition consists of the student pages of *Building Skills in Geography* with answers to exercises annotated.

Organization of the Student Text

Building Skills in Geography is organized around the five fundamental themes in geography identified by the Joint Committee on Geographic Education of the National Council for Geographic Education and the Association of American Geographers in the publication *Guidelines for Geographic Education.* This study, which was completed in 1984, addresses the problem of a continuing lack of geographical understanding on the part of Americans. A 1988 test administered by the Gallup Organization revealed the depth and breadth of the problem. Americans of all age groups ranked sixth. Americans 18 to 24 years of age ranked last. Taking all age groups together, 75 percent of Americans did not know the location of the Persian Gulf. Forty-five percent could not locate Central America. Only one-third knew where Vietnam is, and fewer than half could identify the United Kingdom, France, South Africa, and Japan.

Each unit in *Building Skills in Geography* deals with one of the five fundamental themes: LOCATION, PLACE, HUMAN-ENVIRONMENTAL RELATIONSHIPS, MOVEMENT, and REGIONS. A **Unit Introduction** identifies and exemplifies the theme dealt with in the unit. Each unit also lists objectives and interesting bits of geography trivia. Unit objectives are drawn from the learning outcomes suggested by the Joint Committee in *Guidelines for Geographic Education.*

Each unit is comprised of several related **Lessons** designed to present and practice a specific skill or to develop a specific concept. The skill or concept addressed in each lesson is identified by the lesson objective.

Each unit is followed by a **Unit Review** which tests mastery of the skills presented in the unit. A **Final Review** follows Unit 5.

Vocabulary development is an important aspect of *Building Skills in Geography.* Important terms are listed, with their pronunciations and definitions in each lesson introduction. Each term is printed in dark type in the lesson where it is introduced. Terms are always defined in context upon first use in the text. All vocabulary words are included in the **Glossary** at the end of the book.

Teaching Suggestions

Much of the material in *Building Skills in Geography* is designed to teach students to extract information from graphic sources, such as maps, graphs, and tables. Some students who have difficulty reading may not be able to digest adequately the written material on how to read a map.

Some students may have trouble applying what they have read to an actual map-reading situation. Other students may read explanatory material with ease but stumble when confronted with graphics. The following suggestions are designed to assist such students in gaining the most benefit from this book.

Students with Reading Difficulties

Students with reading difficulties may benefit from a variety of techniques. Lessons can be read aloud to such students, or tape-recorded for individual study. If students have difficulty keeping their place in the text when they are asked to look at an example, have them use a marker—such as a strip of stiff paper—to keep track of the line of text they are reading.

Additional vocabulary development work may also be beneficial prior to each lesson.

Students Who Have Trouble Applying Skills

Students who have trouble applying information from the text to actual map- or graph-reading situations may benefit from oral instruction, either on a one-to-one or small group basis, which allows them to ask questions at any point. Such students may also be asked to develop a simplified set of "rules," or steps to follow, in reading each kind of graphic.

Students with Map-Reading Difficulties

Using maps and graphs requires students to "read" a great many symbols, and much of the language used is mathematical. Students who are able to read text information well but are unable to read graphic materials satisfactorily may need reteaching in interpreting symbols and mathematical language.

Students Who Need Enrichment

Students who need more challenging activities can be assigned more map- or graph-making activities similar to those presented in this book. Students of all ability levels benefit greatly from actual construction activities. Whenever possible, activities should have students write on maps or, using reference materials, make their own maps. Such activities require students to apply all they have learned about direction, distance, symbols, and scale. In addition, such kinesthetic activities tend to reinforce learning.

Additional Map Activities Meet Varying Needs

The additional map activities and outline maps which are a part of this Teacher's Manual are designed to be used with students of varying levels of ability. Students who are having difficulty with any particular skill can be assigned work on an outline map which practices that skill.

Activities to accompany each outline map are listed in ascending order of difficulty. That is, the first activity is designed to be used with students with weaker skills, the next is for students with stronger skills, and so on. Students who need a challenge should be assigned activities from the end of each list.

Additional Map Activities

The following activities are designed to use the reproducible outline maps which follow on pages T4—T14 of this Teacher's Manual. Activities for each map are listed in increasing order of difficulty.

Outline Map 1 (The World)

1. Label the continents and oceans.
2. Label the United States, China, India, Russia, Mexico, Canada, and Australia.
3. Label the equator and the prime meridian.
4. Use latitude and longitude to locate specific countries.
5. Draw in approximate boundaries of time zones and indicate correct relative times.
6. Draw in and label world climate zones.
7. Use data from a source such as a world almanac to show the average population density of each country.

Outline Map 2 (The United States and Canada)

1. Locate and label the following: United States, Canada, Central America, Greenland, Alaska, Pacific Ocean, Atlantic Ocean, Gulf of Mexico, Mississippi River, Missouri River, Ohio River, Caribbean Sea, Hudson Bay, Appalachian Mountains, Rocky Mountains.
2. Locate and label major cities in the region.
3. Prepare a map showing the major natural resources of the region.
4. Prepare a product map showing the major agricultural products of the region.
5. Describe the locations of major cities using latitude and longitude.
6. Show the approximate locations of time zones in the region and indicate the correct relative time in each zone.
7. Prepare a population density map of the region.
8. Prepare a map showing the principal natural regions and landforms.

Outline Map 3 (Latin America and the Caribbean)

1. Label the countries and their capital cities.
2. Indicate direction and distance between pairs of capital cities, such as Brasília—Montevideo; Santiago—Lima; Mexico City—Havana; Cayenne—Quito; La Paz—Managua.
3. Label the following landforms and bodies of water: Isthmus of Panama; Andes Mountains, Amazon River Basin; Yucatan Peninsula; Baja California; Atacama Desert; Pacific Ocean; Atlantic Ocean; Gulf of Mexico; Caribbean Sea; Amazon River; Rio Grande.
4. Use latitude and longitude to describe the locations of major cities.
5. Prepare a product map showing the major natural resources of the region.

6. Prepare a trade map indicating the major trading partners for each country in the region.

Outline Map 4 (Europe)

1. Label the Atlantic Ocean, Mediterranean Sea, Adriatic Sea, Aegean Sea, Baltic Sea, North Sea, Arctic Ocean, Strait of Gibraltar, Bay of Biscay, English Channel, Irish Sea, Danube River, Rhine River, and Seine River.
2. Label the North European Plain, the Alps, the Balkan Peninsula, the Scandinavian Peninsula, the Jutland Peninsula, and the Brittany Peninsula.
3. Label the countries and their capital cities.
4. Determine the distance between pairs of cities, such as Rome—Lisbon; Helsinki—London; Vienna—Paris; Athens—Madrid.
5. Label the prime meridian. Draw in approximate time zones and label them with the correct relative time east and west of the prime meridian.
6. Color the map to show physical relief. Then research and write a paragraph on the subject of why invading armies traditionally marched through the Netherlands and Belgium to attack western Europe.
7. Color the map to show areas that are below sea level. Explain how the Dutch people have created new land from the sea. Discuss how this has affected human-environmental interactions.

Outline Map 5 (Russia and the Independent Republics)

1. Label the countries and their capital cities.
2. Label the Black Sea, and the Caspian Sea, the Aral Sea, the Barents Sea, the Baltic Sea, the Sea of Okhotsk, the Sea of Japan, the Dnieper River, the Volga River, the Ob River, the Yenisey River, and the Lena River.
3. Label the East European Plain, the Kamchatka Peninsula, and the Kuril Islands.
4. Measure the distance between points such as Moscow—Vladivostok; Minsk—Tashkent; and Kiev—Irkutsk.
5. Draw in the approximate boundaries of time zones and label them with the correct relative times. Discuss the problems that having so many time zones in one country creates.
6. Make a product map showing agricultural production in Russia and the Independent republics. Discuss the effect of latitude on agriculture.
7. Make a map of principal transportation routes in Russia and the Independent republics. Explain how the location of Russia has affected the movement of goods by land, sea, and air.

Outline Map 6 (North Africa and Southwest Asia)

1. Label the countries and their capital cities.
2. Label the Atlantic Ocean, Mediterranean Sea, Black Sea, Caspian Sea, Red Sea, Gulf of Aden, Persian Gulf, Gulf of Oman, and Arabian Sea.
3. Measure distances such as: length of Mediterranean Sea from the Strait of Gibraltar to Lebanon; length of the Red Sea; length and breadth of Lebanon, Qatar, and Algeria; and distances between selected capital cities.

4. Make a rainfall map for the region. Discuss how precipitation affects the lives of the people of the region.

5. Make a map of religions of the region. Describe how these religions affect the cultural life of the region.

6. Make a map showing the spread of desertification in the region. Discuss the causes of desertification.

Outline Map 7 (Africa South of the Sahara)

1. Label the countries and their capital cities.

2. Label major geographic features such as the Namib Desert, the Kalahari Desert, the Congo Basin, the Sudd, the Niger Basin, the Serengeti Plain, and the Nubian Desert.

3. Label rivers such as the Niger River, the Congo River, the Blue Nile River, the White Nile River, the Nile River, the Zaire River, the Limpopo River, the Zambezi River, the Orange River, Lake Victoria, Lake Chad, and Lake Tanganyika.

4. Make a product map showing the principal natural resources of the region. Discuss how the region's natural resources and the colonial system have affected the economic development of the region.

Outline Map 8 (South Asia)

1. Label the countries and their capital cities.

2. Label landforms such as the Western and Eastern Ghats, the Sind Plain, the Punjab Plain, the Deccan Plateau, the Thar Desert, the Himalayas, the Hindu Kush, and Mount Everest.

3. Label bodies of water such as the Indian Ocean, the Bay of Bengal, the Krishna River, the Ganges River, the Brahmaputra River, and the Indus River.

4. Make a map showing the prevailing winds in the region during summer and winter. Describe how these winds and geographical features such as the Himalayas affect the lives of people in the region.

5. Make a population density map of the region. Describe how the numbers of people have affected the development of this region.

Outline Map 9 (East Asia)

1. Label the countries and their capital cities.

2. Label major landforms such as the Gobi Desert, the Manchurian Plain, the North China Plain, the Sichuan Basin, the Yunnan Plateau, and the Plateau of Tibet.

3. Label bodies of water such as the South China Sea, the Pacific Ocean, the East China Sea, the Yellow Sea, and the Sea of Japan.

4. Make a relief map showing land that is below 2,000 feet above sea level and land that is above 2,000 feet above sea level. Add symbols showing agriculture in the region. Discuss the relationship between altitude and agriculture.

5. Show the major port cities of the region. Discuss the role of water transportation and location in the spread of cultures in the region.

Outline Map 10 (Southeast Asia)

1. Label the countries and their capital cities.

2. Measure distances between pairs of cities such as Jakarta—Singapore; Rangoon—Ho Chi Minh City; and Manila—Hanoi.

3. Label major islands and island groups such as Java, Sumatra, New Guinea, and the Moluccas.

4. Give the locations in latitude and longitude of cities north and south of the equator, such as Jakarta, Kuala Lumpur, and Phnom Penh.

Outline Map 11 (Antarctica, Australia, and Oceania)

1. Label the continents, countries, and capital cities.

2. Label major island groups such as the Midway Islands, Solomon Islands, Hawaiian Islands, and Marshall Islands.

3. Label Guam, Eniwetok, Bikini, and Wake islands. Research and explain the roles these islands have played in United States history. Describe the importance of location in these roles.

4. Make a precipitation map of Australia. Describe how rainfall amounts affect population patterns.

5. Make a natural resources map of Antarctica. Describe how the region's climate has affected the exploitation of those resources.

Outline Map 1 The World

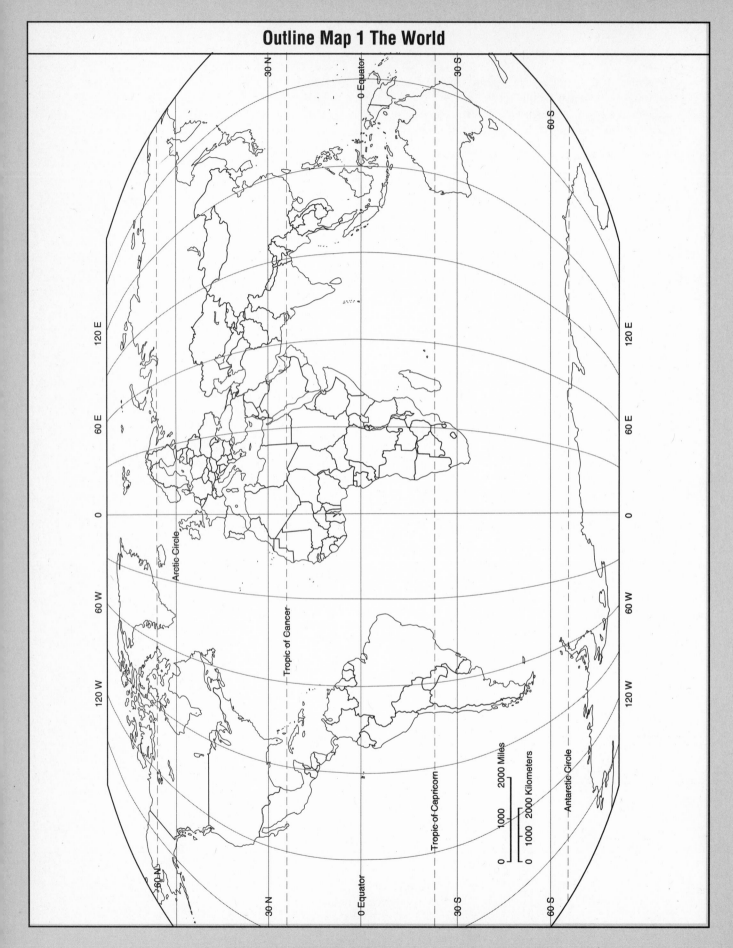

Outline Map 2 United States and Canada

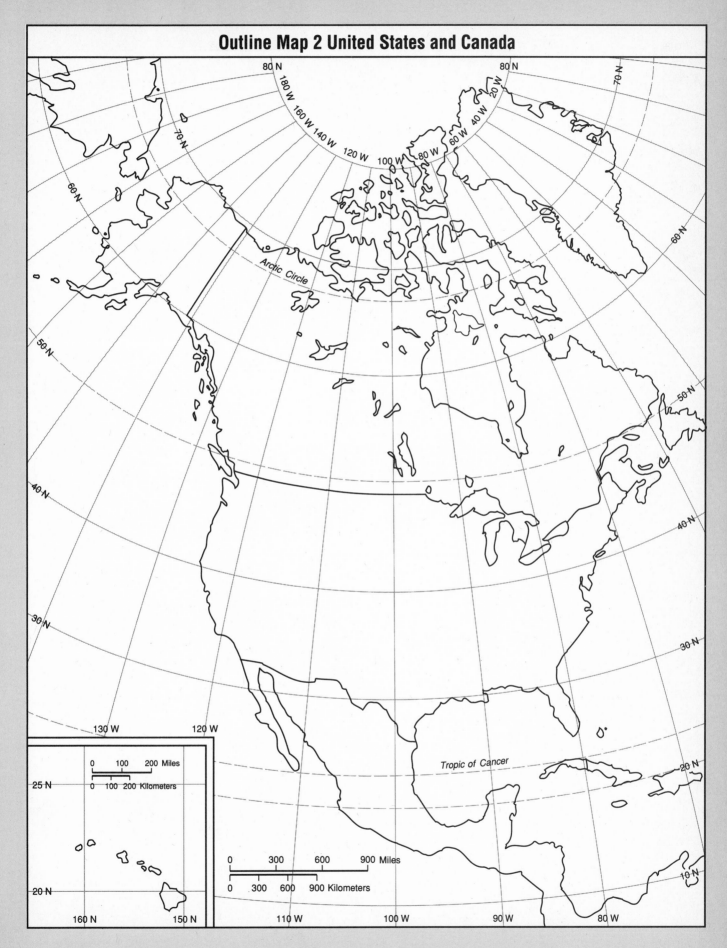

Arctic Circle

Tropic of Cancer

0   100   200 Miles

0   100 200 Kilometers

0   300   600   900 Miles

0   300   600   900 Kilometers

Outline Map 3 Latin America and the Caribbean

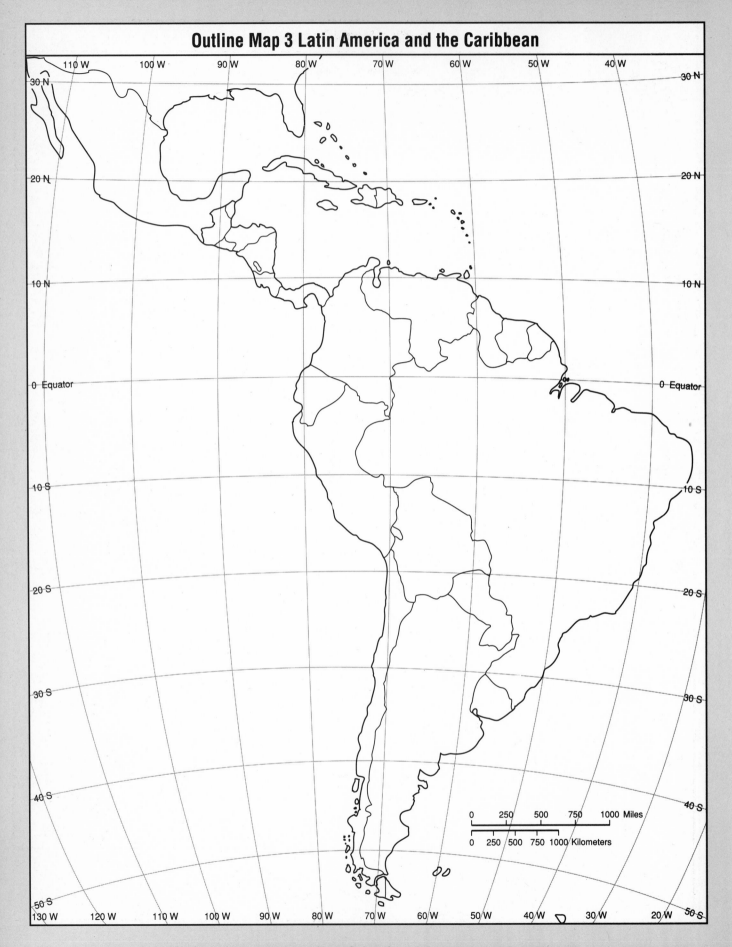

Outline Map 4 Europe

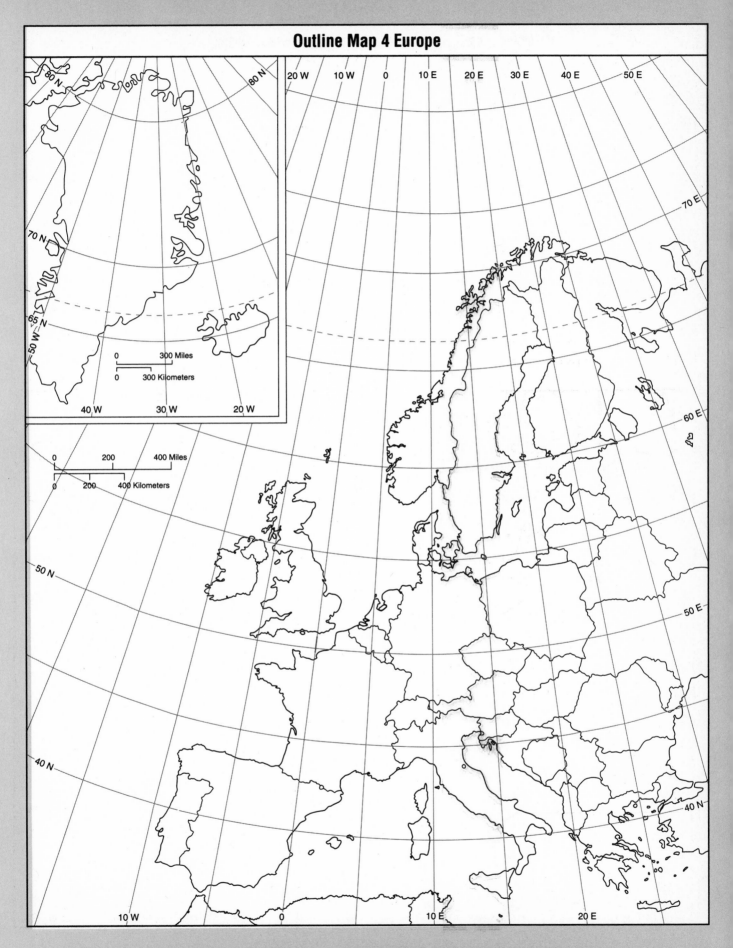

80 N

80 N

70 N

65 N

50 W

40 W   30 W   20 W

0   300 Miles

0   300 Kilometers

20 W   10 W   0   10 E   20 E   30 E   40 E   50 E

70 E

60 E

50 E

50 N

40 N

0   200   400 Miles

0   200   400 Kilometers

50 N

40 N

10 W   0   10 E   20 E

Outline Map 5 Russia and the Independent Republics

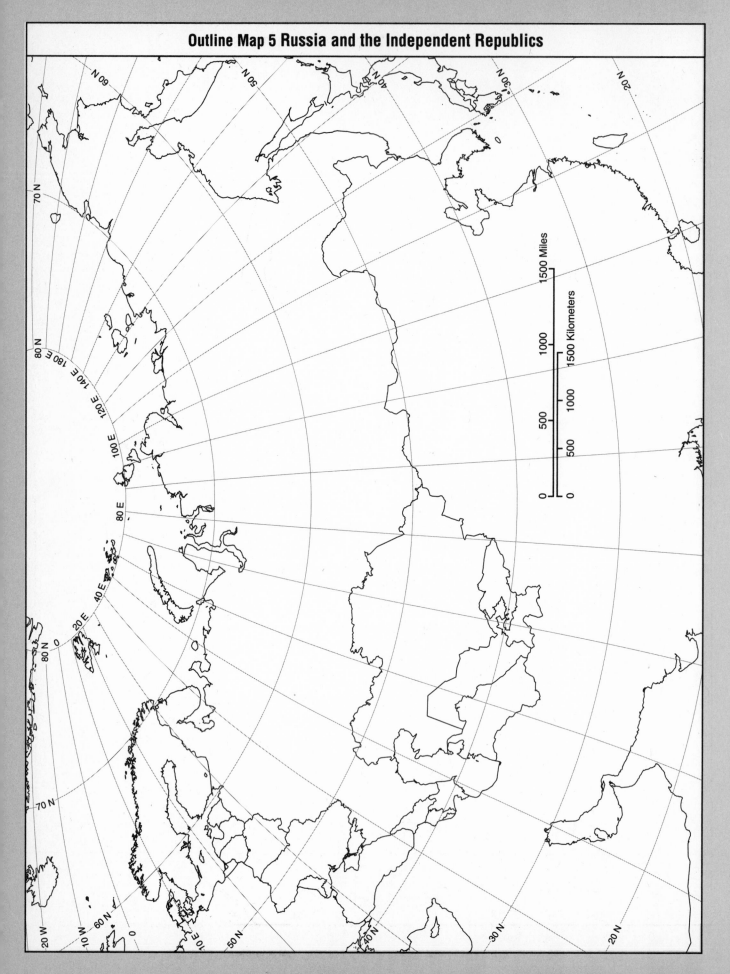

1500 Miles

1500 Kilometers

Outline Map 6 North Africa and Southwest Asia

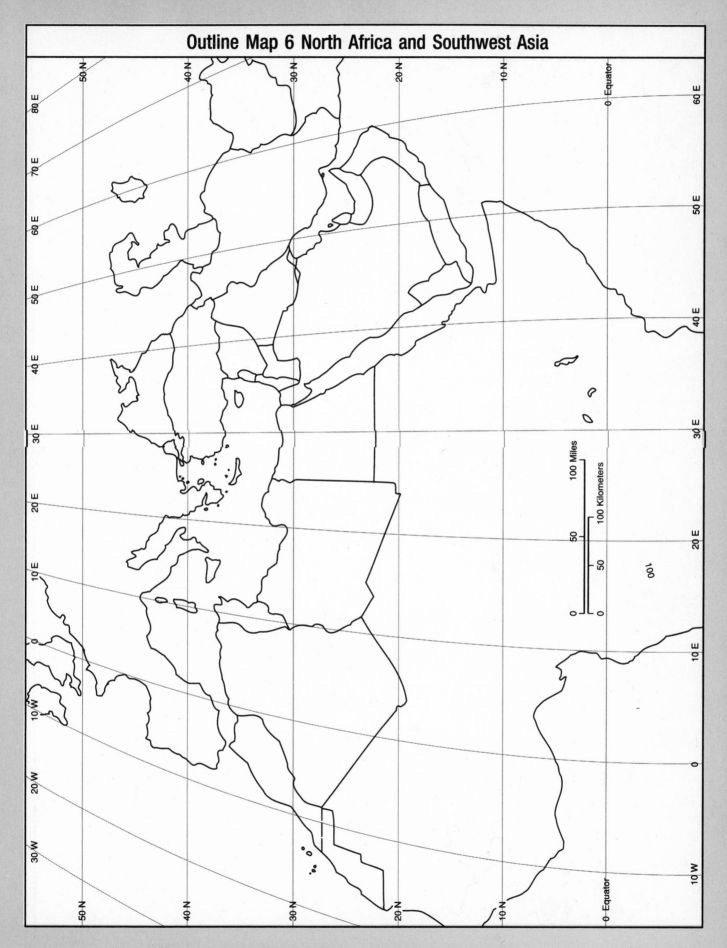

100 Miles

100 Kilometers

100

Outline Map 7 Africa South of the Sahara

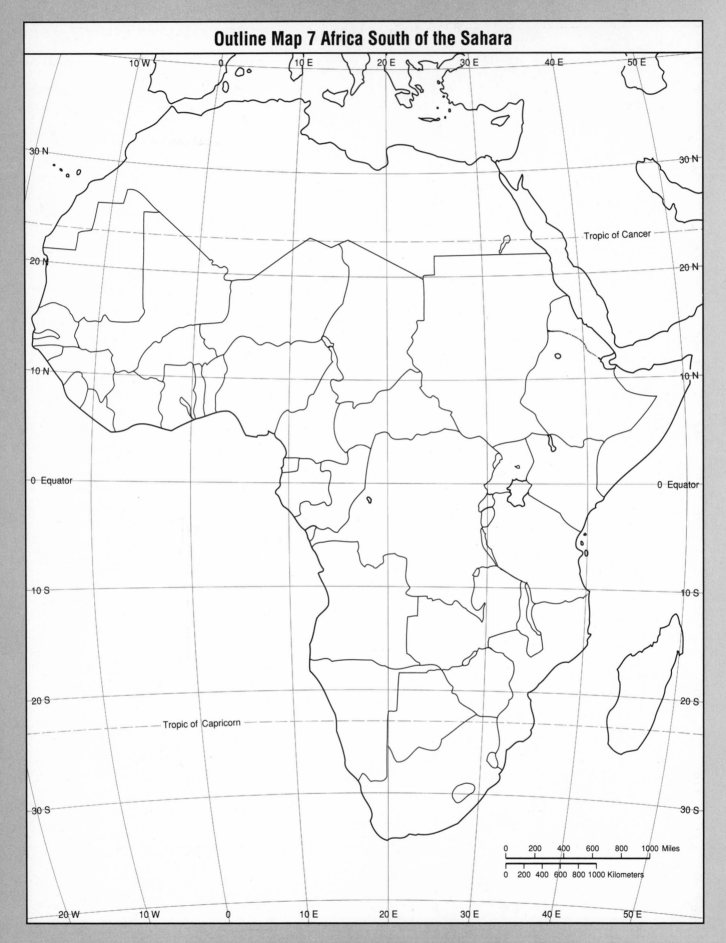

10 W 0 10 E 20 E 30 E 40 E 50 E

30 N 30 N

Tropic of Cancer

20 N 20 N

10 N 10 N

0 Equator 0 Equator

10 S 10 S

20 S 20 S

Tropic of Capricorn

30 S 30 S

0 200 400 600 800 1000 Miles

0 200 400 600 800 1000 Kilometers

20 W 10 W 0 10 E 20 E 30 E 40 E 50 E

Outline Map 8 South Asia

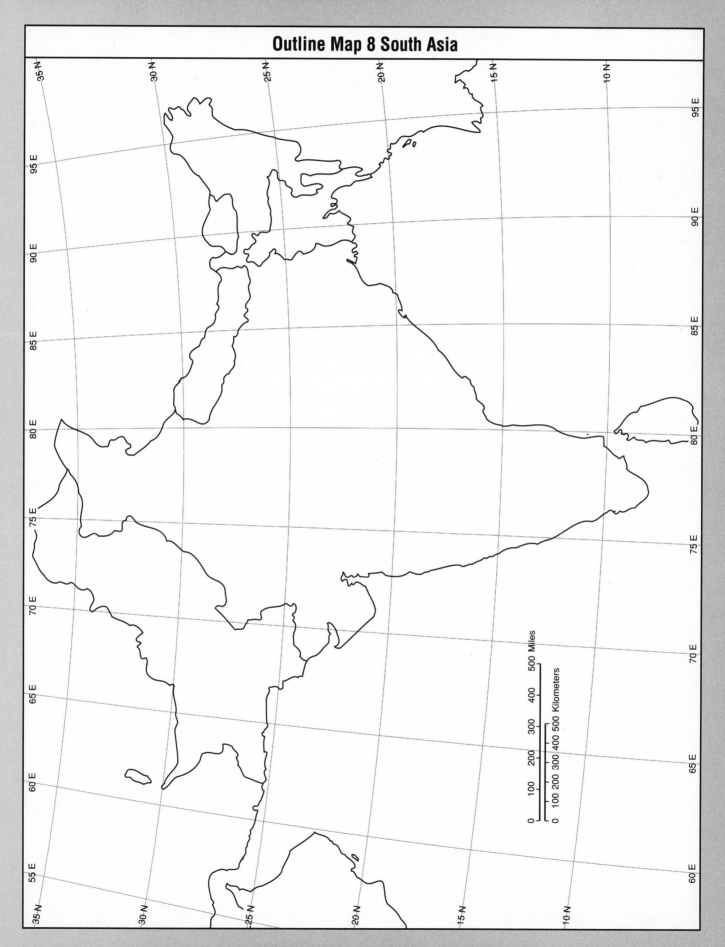

Outline Map 9 East Asia

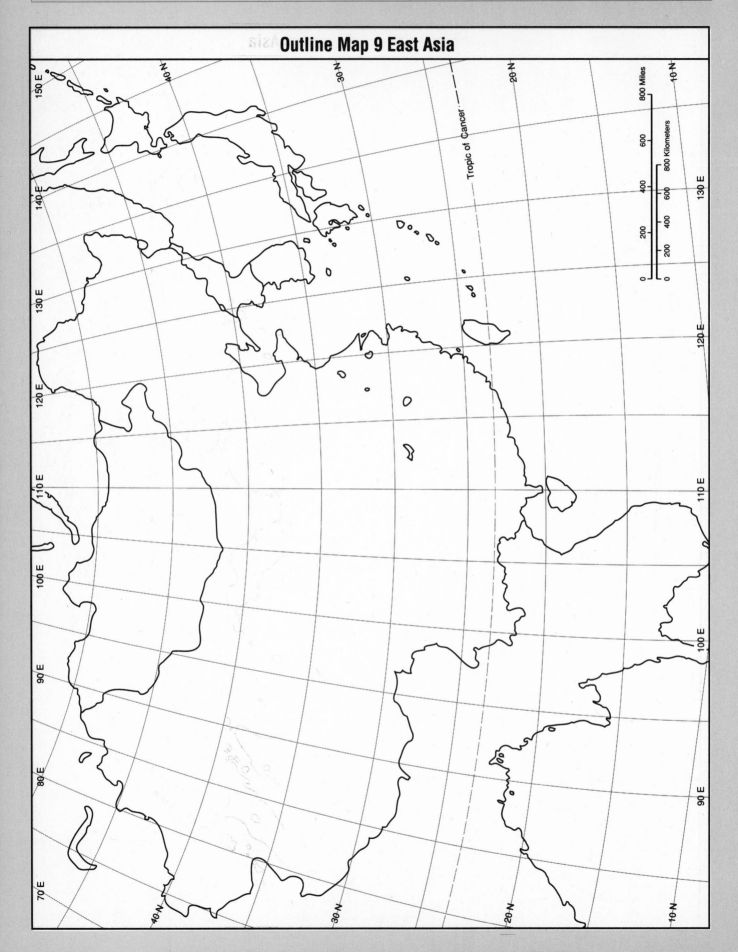

Outline Map 10 Southeast Asia

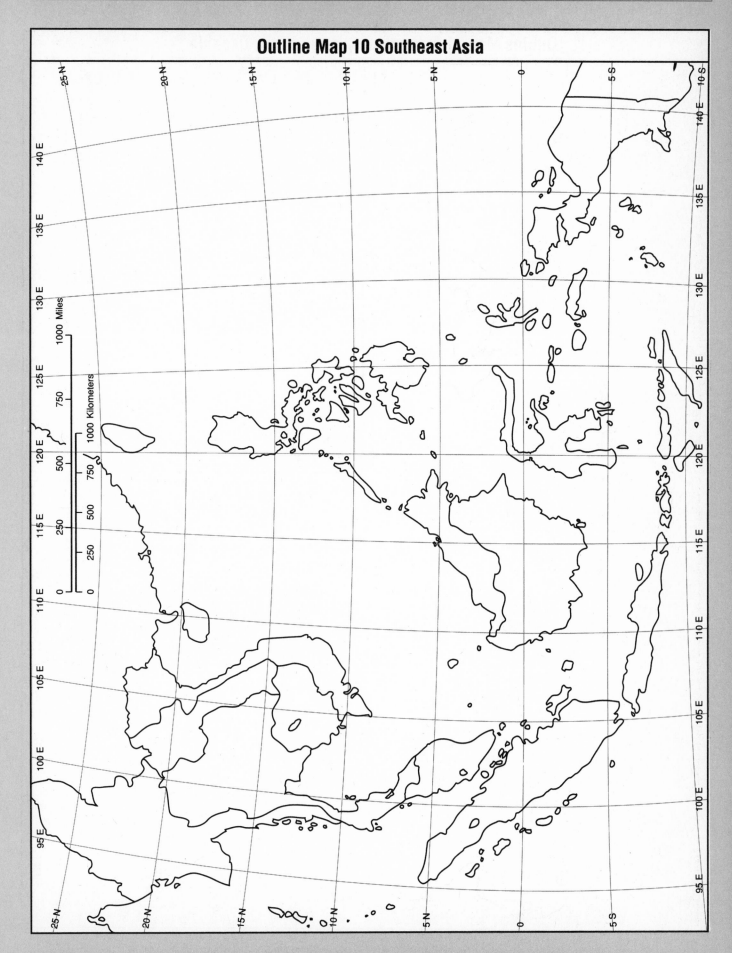

1000 Miles

1000 Kilometers

750

500

250

0

750

500

250

0

Outline Map 11 Antarctica, Australia, and Oceania

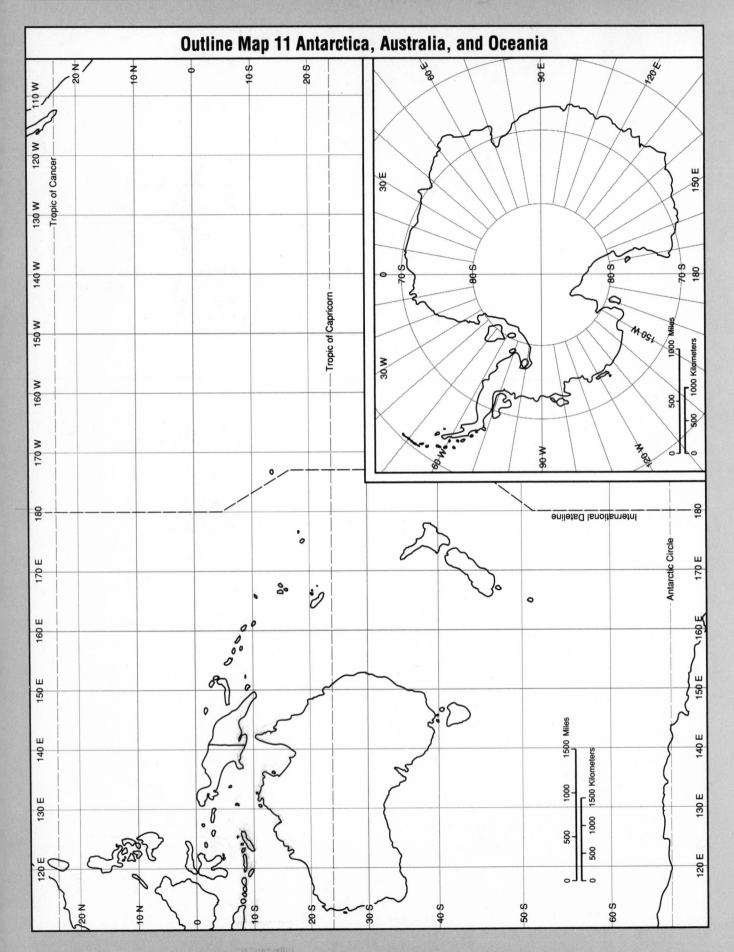

International Dateline

Antarctic Circle

Tropic of Cancer

Tropic of Capricorn

1000 Miles

1000 Kilometers

1500 Miles

1500 Kilometers

Building Skills in Geography

Text-Workbook

Teacher's Annotated Edition

Richard G. Boehm, Ph.D.
Professor of Geography
Department of Geography and Planning
Southwest Texas State University
San Marcos, Texas

Larry L. Bybee
Former Secondary Social Studies Supervisor
Northside Independent School District
San Antonio, Texas

James F. Petersen, Ph.D.
Professor
Department of Geography and Planning
Southwest Texas State University
San Marcos, Texas

GLENCOE
McGraw-Hill

New York, New York   Columbus, Ohio   Mission Hills, California   Peoria, Illinois

Consultants

Ken Boardman
Social Studies Supervisor
Hightstown High School
Hightstown, New Jersey

Ronald R. Romanchek
Instructor
South Oak Cliff High School
Dallas, Texas

Virginia Wilkins
Instructor
Lakewood High School
Lakewood, Colorado

Acknowledgments

The Global Pencil lesson on page 146 is an adaptation of "The International Pencil:
Elementary Level Unit on Global Interdependence," by Lawrence C. Wolken, *Journal of
Geography,* November/December, 1984, pp. 290–293. Used by permission.

Artists

Graphics II
Randy Miyake
Ellen Stern
Dirk Wunderlich

Photographers

p. 6 NASA; p. 31 Tony Freeman/PhotoEdit; p. 48 Tony Freeman/PhotoEdit; p. 92 Alan
Oddie/PhotoEdit; p. 103 D. Donne Bryant Stock Photography; p. 111 Jim Zuckerman/
West Light; p. 112 David M. Doody/The Stock Solution; p. 115 (both) Jeff Schultz; p. 117
Robert Frerck/Odyssey Productions; p. 124 Anna Zuckerman/PhotoEdit; p. 126 Jim
Mendenhall; p. 132 Michele Burgess; p. 140 Michele Burgess; p. 144 David Dobbs;
p. 148 Michele Burgess; p. 152 Anna Zuckerman/PhotoEdit; p. 156 NASA; p. 160
Bettmann Newsphoto/Tim Clary; p. 165 Robert Frerck/Odyssey Productions

Cover Photograph: © Imtek Imagineering/Masterfile

Send all inquiries to:
Glencoe/McGraw-Hill
936 Eastwind Drive
Westerville, Ohio 43081

Printed in the United States of America

ISBN 0-02-823724-2 (Student Text-Workbook)
ISBN 0-02-823725-0 (Teacher's Annotated Edition)
2 3 4 5 6 7 8 9   HESS   00 99 98 97

2 3 4 5 6 7 8 9   HESS   00 99 98 97

Contents

To the Student

D o you live in a city, or in a small town? Or do you live on a farm or a ranch? Has your family lived there a long time or a short time? How does living where you do affect the way you live? For example, do you need special clothes in the winter? Is your favorite food grown nearby, or is it brought in from far away?

All these questions have to do with **geography.** Geography is the study of our home—the earth—and how our lives are affected by it. Almost every detail of our lives is affected by geography. The clothes we wear, the food we eat, the things we do for fun, and the kinds of homes in which we live are all connected to geography.

Ask yourself these five questions about a food in the store where you shop. Where was it grown? What was the place like? How did the people who grew the food live? How was the food brought to your store? In what other places on the earth could that food be grown?

These five questions deal with the very heart of geography. As you look for answers to these questions, you are studying geography.

When we study a place on the earth, we ask five questions which are very similar to the five above. We ask (1) where a place is, (2) what the place is like, (3) how the people there live, (4) how the people in that place interact with people in other places, and (5) how that place is like other places on the earth. These questions deal with the five basic themes of geography. These themes are **location, place, human-environmental relations, movement,** and **regions.**

This book is divided into five units. Each unit is organized around one of the five basic themes of geography. Each unit introduction will explain more fully the theme being dealt with in that unit.

Each unit introduction is followed by a series of short lessons that deal with the topic of that unit. Numerous maps, graphs, and tables will be used to present information. You will receive a great deal of instruction and practice in reading these special ways of presenting information. A **Unit Review** checks your

understanding of the important concepts and information presented in each unit. A **Final Review** at the end of the book is an overall check on your learning.

Vocabulary study is an important part of this book. Geographers use many special words to describe the earth and its peoples. You need to know these words in order to understand the world in which you live. The terms you should know after completing the lesson are listed at the start of each lesson. All the words are also listed in the **Glossary** at the end of the book, so that you may easily look them up at any time. The Glossary tells how to pronounce each word and gives the meaning of the word.

As you progress through the book, you will learn some of the skills you need to learn about places on the earth. You will learn some of the language of geography. You will learn how to read the maps, charts, and tables that geographers often use to present information about the earth and its peoples. You will study examples of how people interact with each other and with their environment. And you will learn how to organize your study of the earth by areas that are alike in some way.

Almost every day in the newspaper or on television, we learn of some place of which we might have never heard before. Often we find that in some way our lives are affected by that place. Events in places such as Bihac, Bosnia-Herzegovina; the Gaza Strip; and Mogadishu, Somalia impact our lives as part of a global community. We may know someone from those parts of the world. We at least have read about and heard about their misfortunes. Perhaps we have donated money to help buy food for war-torn areas, or to help establish homes for orphans.

All of us, everywhere, are affected in some way by what happens everywhere else. We study geography to help us deal with the things that affect our lives. The skills you will learn in this book can help you make better decisions about where and how you will live. In that sense, this book is designed to help you gain more control over your own life.

UNIT 1 Location: Position on the Earth's Surface

OBJECTIVES

After completing this unit, you will be able to:

- describe locations in absolute and relative terms,

- locate places on a map using latitude and longitude,

- locate major landmasses and bodies of water,

- read map symbols, legends, and scales,

- compare different types of maps and map projections.

A simple human action can have a complicated effect. On April 26, 1986, in a nuclear power plant at Chernobyl, Ukraine, a technician turned a valve. Due to mistakes that had already been made, the nuclear reactor began to build up power far too rapidly. Within three seconds, the reactor was operating at almost 500 times the top power it was designed to reach. One minute later, the reactor exploded and caught on fire. Deadly radioactivity spewed into the air. Winds began carrying the particles northwestward into Europe and eventually around the world. Dozens of people were killed. Hundreds more became very ill.

When news of the disaster became known, most people's first question was, "Where is Chernobyl?" That is, they wanted to know exactly where on the face of the earth it was located. The nuclear reactor at Chernobyl takes up one particular spot of ground. This is called its **absolute location.**

Within days after the disaster, news reports began to carry stories about the spread of the radioactive cloud from the explosion. Poland, Sweden, England—all reported increases in radiation. Farmers across Europe were forced to destroy crops covered with radioactive dust. People were afraid to drink the water. Now people were asking a different question about the location of Chernobyl: "Where is it *from here?*" People needed to know in what direction, and how far, Chernobyl was from their own location. This is called **relative location.** By knowing your location in relation to Chernobyl, you could tell whether the wind was carrying the radiation closer or farther away.

Geographers use latitude and longitude to describe absolute locations. Lessons in this unit will teach you how to use latitude and longitude. Geographers also use absolute location to make maps of the earth's surface. This unit will teach you how to read several kinds of maps. You will also learn about relative location and its importance to you.

Location: position on the earth's surface is one of the five basic themes of geography. While location does not determine how we live, it does influence our lives. Imagine how different your life would be if you lived at the North Pole or in the middle of the Pacific Ocean.

Tallest Mountain

For many years Mount Everest in Nepal has been recognized as the world's tallest mountain. Everest, which is part of the Himalaya mountain chain, is 29,028 feet above sea level at its peak. Now, however, new measuring methods using satellites suggest that Everest might be the second-tallest mountain on Earth.

The satellite's measurements indicate that a mountain in Pakistan called K2 might be taller than Everest. Geographers, geologists, and other scientists are studying this new information to determine which peak really is the tallest mountain in the world.

Spaceship Earth

As you read this you are traveling through space at a speed of over 66,000 miles an hour. In fact, we all are. The earth travels around the sun at an average speed of 18 and 1/2 miles a second or over 66,000 miles an hour.

If that figure makes you feel dizzy, you probably don't even want to think about this one. The earth rotates on its axis at a speed of over 1,000 miles per hour. So, even when you think that you're doing nothing, you are not. You're spinning and traveling on this spaceship Earth.

7

OBJECTIVE

Identify direction and distance information from maps

TERMS TO KNOW

compass rose (KAHM·puhs rohz)—symbol used on a map to show directions

north arrow (north EHR·oh)—symbol used on a map to show directions

scale (skayl)—symbol used on a map to show distance

Have you ever drawn a map in the dirt to show someone where you live? Such drawings were some of the earliest maps. Other early maps were made of sticks tied together, or pieces of wood sewn to a piece of sealskin. People have used maps for thousands of years to show *where* places are, *how far* it is from one place to another, and the *direction* to travel to get from here to there.

Maps are important tools. Maps tell us where to catch a bus and where that bus will take us. Maps help us find a friend's house in a part of town that is new to us. Maps help us plan vacation trips. They help us learn more about the town or state to which we are moving.

Direction

Direction is one of the most important things we can learn from a map. You use direction every day—left, right, forward, back, up, down. But these directions depend on where you are and which way you are facing. Maps use the directions north, south, east, and west. These directions do not change. North is always toward the North Pole of the earth. If you stand facing the North Pole, east will be to your right. West will be to your left. South will be behind you.

Usually, north will be at the top of a map. However, this is not always true. You must check to be sure. Mapmakers use a **compass rose** or a **north arrow** to show directions. If there is no compass rose or north arrow, or other symbol to indicate direction, north should be at the top of the map.

Look at the examples. Find north, south, east, and west on Example 1–1. Turn your book so that north on the compass rose points north. Face north yourself. Now east is to your right, west is to your left, and south is behind you.

Look straight north. Hold your right hand straight out to the side. In what direction are you pointing? You are correct if you said *east*. Now turn your head just halfway toward your right arm. You are no longer looking north. But you are not looking east, either. You are looking *northeast*. Look at Example 1–2. Find northeast, southeast, northwest, and southwest.

Notice that Example 1–3 is just an arrow with its point labeled *N*. The *N* stands for north. The arrow points north. When you see a north arrow, remember that east is to the right, west is to the left, and south is opposite north.

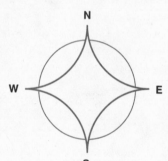

Example 1–1

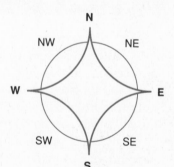

Example 1–2

Example 1–3

8

Distance

If you make a drawing of a person, you will probably not make the picture as large as the person. That would take a piece of paper the same size as the person. A map is a drawing of a part of the earth. A map as big as the earth would be too large to put in your pocket and carry with you across Africa! Maps are drawn so that a certain distance on the map represents a much larger distance on the earth. This makes it possible to show the whole earth on a piece of paper the size of this page.

Not all maps are the size of this page, of course. Your classroom may have maps hanging on the wall. These maps are much larger than the ones in your book. But they both show the same earth. Maps have a **scale** to tell you what distance on the earth is represented by a certain distance on the map.

Using Map Scales

Here are some examples of map scales. Notice that all lines are the same length, but that each line represents a different distance on the earth. Also notice that the same scale can tell how many miles *and* how many kilometers each distance on the map stands for.

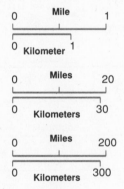

Using the scale to measure distances between places on a map is easy. Use a piece of paper. Put the edge of the paper between the two points you wish to measure. Make a mark on the paper at each point. Then put the piece of paper on the map scale with one mark at zero. Note where the other point falls on the scale. This gives you the distance.

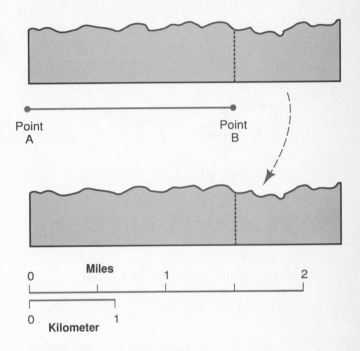

If the scale is not long enough, mark where it ends on the paper. Then slide the paper to the left to line up the new mark with zero. Do this as many times as necessary. Then multiply the number of spaces between marks times the distance each length of scale represents. For example, if the scale represents 100 miles, and you marked off three spaces, multiply 3 times 100. The distance between the two points on the map is 300 miles.

Using Your Skills

 Place each phrase in the box under the correct heading.

| helps you find directions on a map | helps you find distances on a map |
| can be marked in miles | may be marked N, S, E, W |

Compass Rose

1. helps you find directions on a map

2. may be marked N, S, E, W

Scale

3. can be marked in miles

4. helps you find distances on a map

B   Fill in the missing directions on these compass roses. Notice that north is not always in the same place.

1.
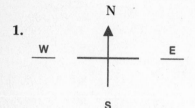

2.
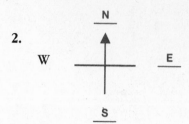

3.
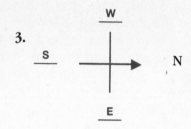

4.
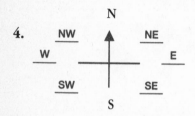

5.
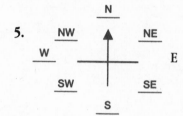

6.

C   Use what you have learned about direction and distance to answer the questions about Map 1–1.

1. What part of this map shows direction?   the compass rose

2. What part of this map shows distance?   the scale

3. If you were in Kansas, in which direction would you have to travel to reach each of the states below? Use directions such as *southeast* when necessary.

a. South Dakota   north   b. Virginia   east

c. Utah   west   d. Texas   south

e. Washington   northwest   f. Florida   southeast

g. New Mexico   southwest   h. Michigan   northeast

4. How many miles does the full length of the scale on the map stand for?   400

5. About how many miles is it from east to west across Colorado?   400

6. How would you measure a distance on the map that is longer than the scale?   Place a piece of paper on the map and mark each end point of the distance to be measured. Place one end point on the scale at zero. Put a mark at the end of the scale. Slide the paper over so the mark lines up with zero. Keep doing this until you reach the end point. Then multiply the number of spaces times the distance the scale stands for.

7. About how many miles is it from north to south across Texas at its widest point?

800

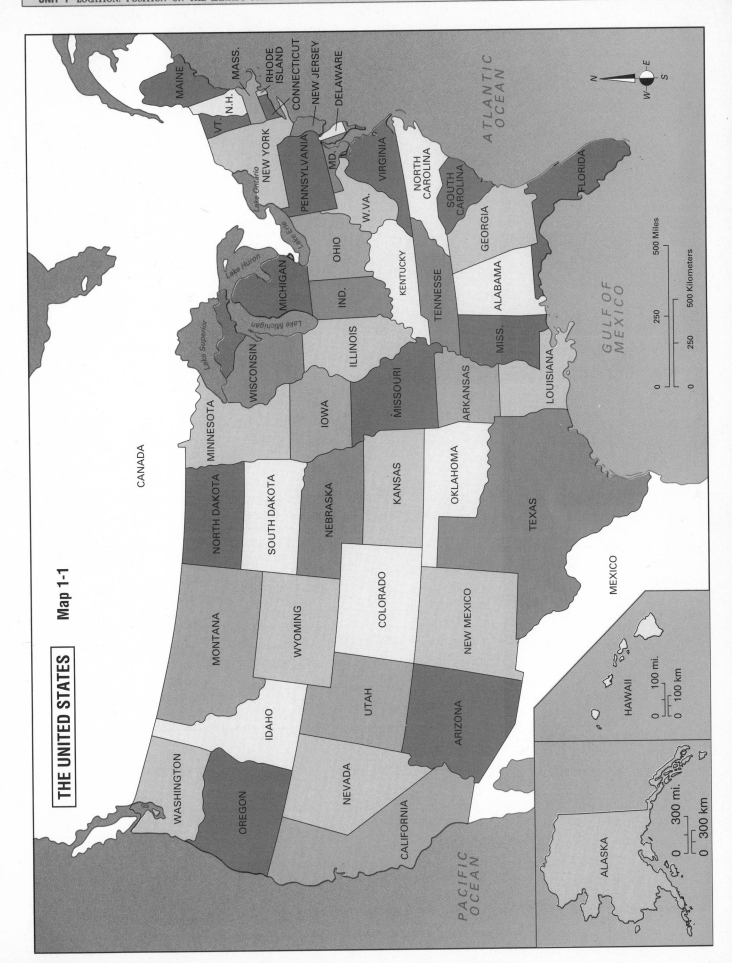

THE UNITED STATES   Map 1-1

Describe locations in terms of relationships with other locations

interdependence (in·ter·dee·PEN·duhns)—dependence upon one another

relative location (REL·uh·tiv loh·KAY·shuhn)—location compared to the location of other places on earth

Have you ever been lost? Or have you just not been sure about how to get somewhere you wanted to go?

From the one spot you stand on the earth, you are different directions and distances from many other spots on the earth. You may be 22 miles south of your home. At the same time you may be 150 miles northeast of the capital of your state. You may also be three feet from the front door of your favorite pizza place. Your location can, in fact, be compared to the location of any other spot on earth. This is called **relative location.**

Relative location affects your life in countless ways. If you live 8 miles from school, you must wake up earlier each morning than someone who lives 8 blocks away. If you live 2,500 miles from the nearest volcano, you will be much less concerned about its latest eruption than someone who lives in the valley below it.

The story of Houston, Texas, is a good example of the importance of relative location. One of the greatest oil strikes in history took place near Houston in 1901. The Spindletop Field was the first great oil discovery. Others in Texas followed. Within a few years Houston, Texas, was an important center for the oil industry. Why? Because Houston's relative location was near the early oil fields. It was also located near the Gulf of Mexico. This made it possible to ship oil and equipment by water. Many oil companies built plants near Houston to make products from oil. These products were then shipped by water. Houston became one of the largest ports in the United States.

Global Interdependence

Relative location is important because we depend on people in other places for things we need. Depending on other people is called **interdependence.** We depend on them, and they depend on us.

Interdependence links us together in many ways. For example, the United States buys from other countries much of the oil that runs its cars and factories. A great deal of this oil comes from Southwest Asia. This is why the United States is so interested in wars and other events in Southwest Asia. Because of the relative location of the oil fields to Southwest Asia, a war there could cut off the supply of oil to the United States.

Using Your Skills

A Use Map 1–2 to decide whether each statement about relative location is true or false. Write *T* if the statement is true. Write *F* if the statement is false.

_____T_____ **1.** Panama has water to the north and south.

_____F_____ **2.** The country of Colombia is located to the west of Panama.

_____T_____ **3.** The Caribbean Sea is located to the north of Panama.

_____T_____ **4.** The Canal Zone is located in the central part of Panama.

_____T_____ **5.** When a ship enters the Panama Canal at Colón, it is northwest of the other end of the canal at Panama City.

_____T_____ **6.** A person living in Chitré, Panama, will see the sun come up over the Pacific Ocean.

_____F_____ **7.** According to this map, all parts of the Caribbean Sea are east of the Pacific Ocean.

_____F_____ **8.** The city of Rio Hato is about 300 kilometers west of La Palma.

_____F_____ **9.** Las Tablas is located about 100 kilometers southeast of Panama City.

_____T_____ **10.** The Isla del Rey is located about 200 kilometers northeast of the Isla de Coiba.

_____F_____ **11.** The Golfo De Los Mosquitos is located in the Pacific Ocean.

_____F_____ **12.** The city of Las Tablas is located about 55 kilometers north of Rio Hata.

_____T_____ **13.** Both La Palma and Puerto Armuelles are port cities on the Pacific Ocean.

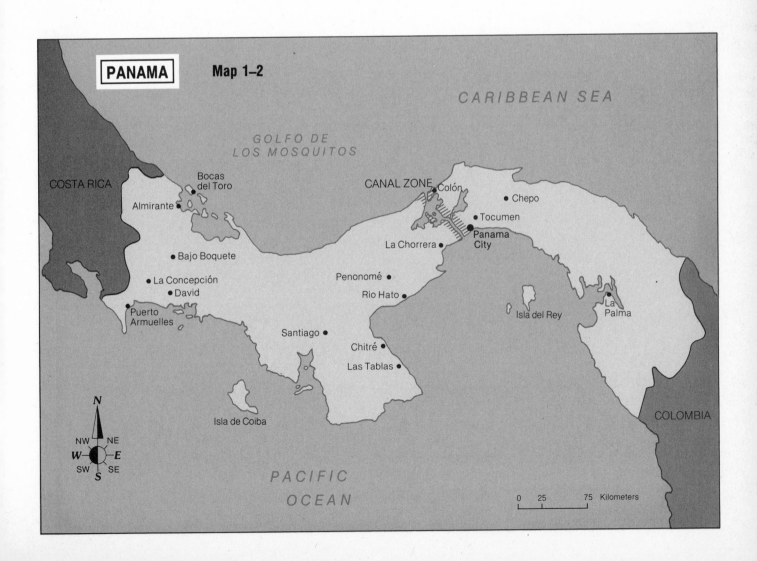

B   Use Map 1–3 to answer the questions about relative location. Use directions such as *northeast* where necessary.

1. Where is the United States located on this map?   The United States is located in the middle of the map near the top.

2. What country is to the north of the United States?   Canada

3. What country is to the south of the United States?   Mexico

4. What direction is South Africa from the United States?   southeast

5. What direction is Australia from the United States?   southwest

6. In what direction would you travel in going from Japan to the United States?   east

7. In what direction would you travel in going from India to the United Kingdom?   northwest

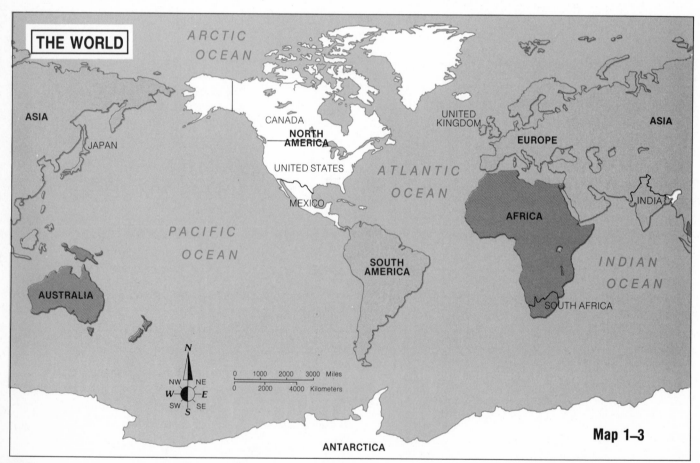

Map 1–3

C   Use maps of your state, the United States, and the world to answer these questions.

1. In what direction, and how far, is the capital city of your state from your home?   Answers will vary.

2. In what direction, and how far, is Washington, D.C., from your home?   Answers will vary.

3. In what direction, and how far, is the Atlantic Ocean from your home?   Answers will vary.

4. In what direction, and how far, is the Pacific Ocean from your home?   Answers will vary.

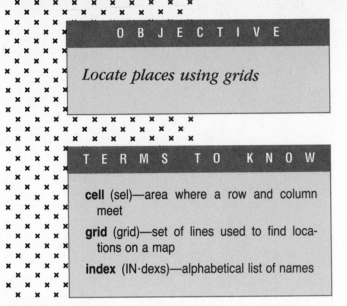

Lesson 3 Locating Places Using a Grid

O B J E C T I V E

Locate places using grids

T E R M S   T O   K N O W

cell (sel)—area where a row and column meet

grid (grid)—set of lines used to find locations on a map

index (IN·dexs)—alphabetical list of names

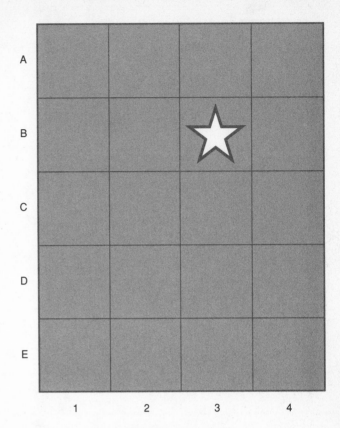

Example 1–4: An Alpha-Numeric Grid

I magine that you have just landed on this planet. Someone hands you a dictionary and asks you to look up the meaning of *word.* You did not know that people on this planet still used books. You thought that anytime anyone wanted to know something, they just talked to their computer. You know something about English, so you know that the letter *w* stands for the sound at the beginning of *word.* But your ship had to leave before you learned about vowel sounds. What letter comes next? Is *word* spelled wird, wurd, werd, word, or whirred? It's hard to look up a word you can't spell.

Reading a map can be a little like that. Suppose your teacher asks you to find the location of Wroclaw, Poland, on a map. You look at a map of Poland. It seems as though every town's name has a bunch of w's, c's, and z's. You need something that will tell you *about* where Wroclaw is located.

Using Map Grids

The something you need is called a **grid.** A grid is a set of lines used to identify locations on a map. Letters and numbers around the edges of the map label the areas marked off by the lines. Look at Example 1–4.

Place your left index finger on the letter *B* on the left side of the grid. Place your right index finger on the number *3* at the bottom of the grid. Move your left finger straight across and your right finger straight up until they meet. There should be a star at your fingertips.

The four spaces to the right of the letter *B* are in a row. We call this row B. The four spaces above the number *3* are in a column. We call this column 3.

The area where a row and a column meet is called a **cell.** Notice that only one cell can be at the area where row B and column 3 meet. We call this cell B-3.

Practice using the grid. Draw another star in cell C-2. Draw a circle in cell A-4. Write your name in cell D-1. You should be able to draw a straight line through all four cells with something in them.

Using a Grid Index

Mapmakers often use a grid to help us find places on maps. The grid is used with an **index.** The names of places on the map are listed in alphabetical order in the index. Following each name is the letter and number of the cell in which that place can be found.

Look at Map 1–4 and the index. Notice that the index is not complete. Fill in the name of the missing city for each cell number.

Map 1–4

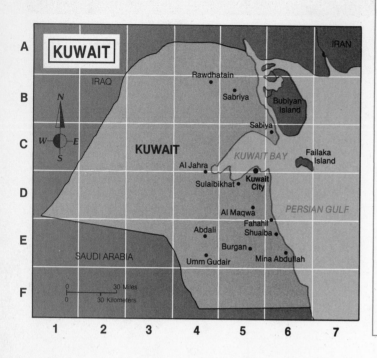

Index

Abdali E-4
Al Jahra D-4
Al Maqwa D-5
Burgan E-5
Fahahil E-6
Kuwait City C-5, D-5
Mina Abdullah E-6

_____ Rawdhatain _____ B-4

_____ Sabiya _____ C-6

_____ Sabriya _____ B-5

_____ Shuaiba _____ E-6

_____ Sulaibikhat _____ D-5

_____ Umm Gudair _____ E-4

Using Your Skills

A **Match each term at left with its meaning.**

___c___ **1.** cell

___d___ **2.** row

___e___ **3.** column

___b___ **4.** grid

___a___ **5.** index

a. an alphabetical list of places on a map, with cell numbers

b. a set of lines used to identify locations on a map

c. the space where a row and column meet

d. a set of spaces which go across a map

e. a set of spaces which go up and down a map

B **Use Map 1–5 to answer these questions.**

1. What is located in cell B-3?   _____Planetarium_____

2. What is located in cell C-2?   _____Hyde Park_____

3. What is located in cells F-2, F-3, F-4, F-5, E-5, D-5, and C-5?   _____the Thames River_____

4. Complete the following index for the map of London. Remember that all names in an index are in alphabetical order. If there is more than one possible answer for a cell, see which answer will fit in alphabetical order.

Map 1–5

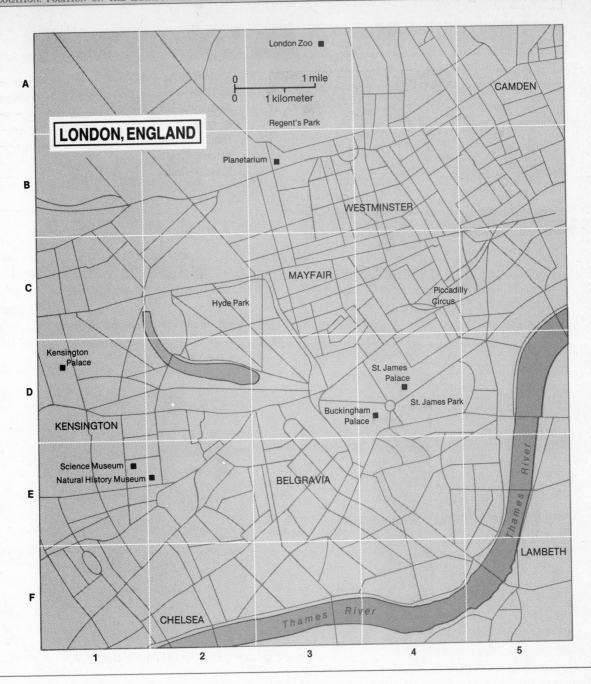

Index

Lesson 4 Introduction to Latitude and Longitude

What they needed was a grid system that covered the entire earth. You know that a grid is made up of two sets of lines. These lines cross each other. A grid system that covered the whole earth would let anyone find any location on earth. We have such a grid today. We call it **latitude** and **longitude.**

Using Latitude and Longitude

Latitude lines, called parallels, run east and west. Longitude lines, called meridians, run north and south. Latitude and longitude are measured in degrees. The earth is a sphere. It is 360 degrees around a sphere. Each degree of latitude or longitude is 1/360th of the distance around the earth. The symbol for degree is °.

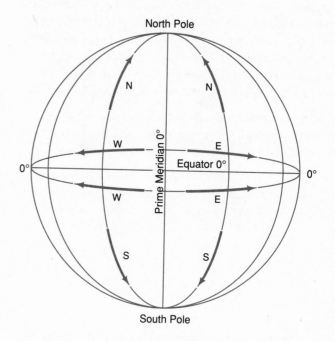

D o you know how ships measured their speed long ago? Do you know why a ship's speed is given today in knots rather than miles per hour or kilometers per hour?

Long ago, each ship carried a piece of wood fastened to a rope. The rope had knots tied in it. Each knot was a certain distance from the next. To measure the ship's speed, the piece of wood was thrown overboard. It pulled the rope out behind it. The faster the ship was going, the faster the rope went out. Someone counted how many knots passed over the side of the ship in a certain length of time. If seven knots were pulled out, the ship was said to be traveling at a speed of seven knots. Today, one knot is about 1.15 miles per hour.

Ships of long ago had to keep track of their speed on long voyages because they had no other way to tell how far they had traveled. Ships often became lost. A storm might blow them far away from where they wanted to go.

What people needed was a way to tell exactly where they were on the earth's surface—their absolute location. They also needed to be able to find their way to any other absolute location.

The starting point for measuring degrees of latitude is the **equator.** The equator is a line of latitude. It divides the earth into two equal parts. The equator runs east and west all the way around the world, halfway between the North and South poles. We say that the equator is at zero degrees (0°) latitude. When we give the latitude of a place, we must state whether the place is north or south of the equator. For example, the North Pole is at 90° north latitude. If we said only that a place was at 90° latitude, we would not know if the place was the North Pole or the South Pole.

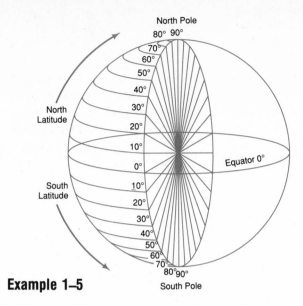

Example 1–5

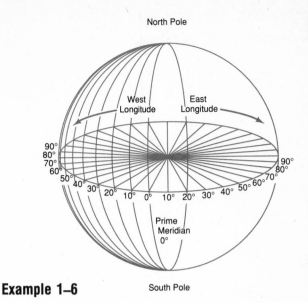

Example 1–6

The starting point for measuring longitude is called the **prime meridian.** Meridian is another name for a longitude line. The earth does not have an east pole and a west pole. Therefore, some point had to be chosen as the starting point for measuring longitude. A place in England, Greenwich, was chosen. All longitude is measured from a line running from the North and South poles through that place in England.

The prime meridian is at 0° longitude. When we give the longitude of a place, we must state whether the place is east or west of the prime meridian.

Lines of latitude run all the way around the earth, but lines of longitude do not. Halfway around the earth from the prime meridian is the line of longitude

marked 180°. This line is the ending point for measuring longitude. The area east of the prime meridian and 180° is east longitude. The area west of the prime meridian and 180° is west longitude. The United States is in west longitude.

Look at Example 1–5 of latitude lines. Find the equator. Now find the line 10° north of the equator. Notice that the angle between the equator, the center of the earth, and this line is 10°.

Now look at Example 1–6 of longitude lines. Find the prime meridian. Now find the line 10° west of the prime meridian. Notice that the angle between the prime meridian, the center of the earth, and this line is 10°.

Using Your Skills

A   **Explain the meaning of each of the following.**

1. degree   A degree is 1/360th of the distance around the earth. Latitude and longitude are measured in degrees.

2. latitude   Latitude is a measure of distance north or south of the equator.

3. longitude   Longitude is a measure of distance east or west of the prime meridian.

4. equator   The equator is the zero degree line of latitude. It runs east and west around the earth halfway between the North and South poles.

5. prime meridian   The prime meridian is the zero degree line of longitude. It runs from the North and South poles through a point in England.

B   **Follow the directions to complete Map 1–6.**

1. Find the line of latitude which is the equator. Write *equator* on the line.

2. Find the line of longitude which is the prime meridian. Write *prime meridian* on the line.

3. The lines of latitude and longitude shown on the map are spaced 30° apart. Find the first latitude line north of the equator. Label the line 30°N. Find the first latitude line south of the equator. Label the line 30°S. Now label the rest of the latitude lines correctly.

4. Find the first longitude line east of the prime meridian. Label the line 30°E. Find the first longitude line west of the prime meridian. Label the line 30°W. Now label the rest of the longitude lines correctly.

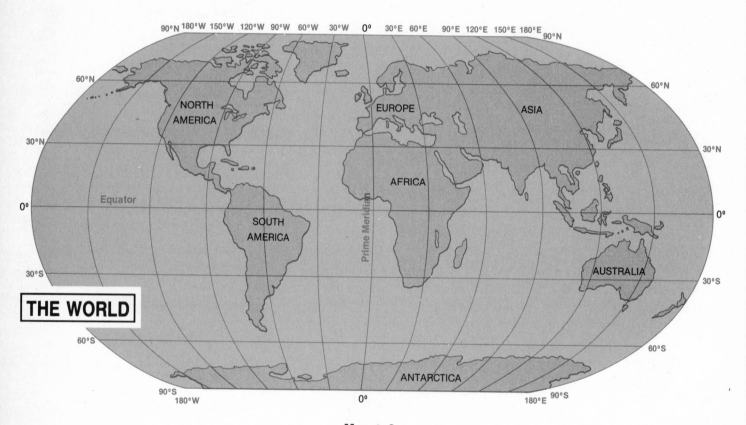

Map 1–6

OBJECTIVE

Locate places using latitude and longitude

Finding places using latitude and longitude is just like using a grid. Look at Map 1–7 of Colorado. Notice that each degree of latitude and longitude is shown. Find the line for 40°N latitude. What city is located on that line? What line of longitude is closest to Boulder? We say that Boulder is located at about 40°N latitude, 105°W longitude. Latitude is always written first.

Now look at Delta. What line of latitude is closest to Delta? What line of longitude runs very near Delta? We say that Delta is located at about 39°N latitude, 108°W longitude.

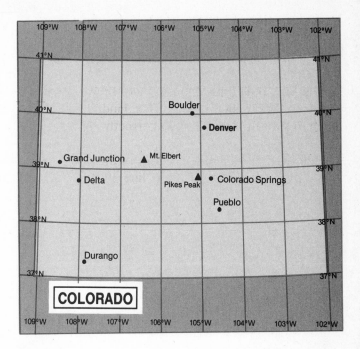

Map 1–7

Using Your Skills

A   **Answer these questions about Map 1–7.**

1. Which line of latitude runs near Pikes Peak? ___39°N___

2. Which line of longitude runs near Pikes Peak? ___105°W___

3. Write the location of Pikes Peak using latitude and longitude. ___39°N latitude, 105°W longitude___

4. Which line of latitude runs nearest Pueblo? ___38°N___

5. Which line of longitude is closest to Pueblo? ___105°W___

6. Write the location of Pueblo using latitude and longitude. ___38°N latitude, 105°W longitude___

7. What city is near 39°N latitude, 105°W longitude? ___Colorado Springs___

8. What city is located at about 39°N latitude, midway between 108° and 109°W longitude?

___Grand Junction___

9. Which line of longitude is at Colorado's eastern border? ___102°W___

10. Which line of latitude runs nearest Grand Junction? ___39°N___

21

B   Use Map 1–8 of Africa to answer the following questions. Be sure always to begin counting degrees of latitude from the equator and degrees of longitude from the prime meridian. Also, be sure to notice that the number of degrees gets larger as you go away from the zero lines.

Map 1–8

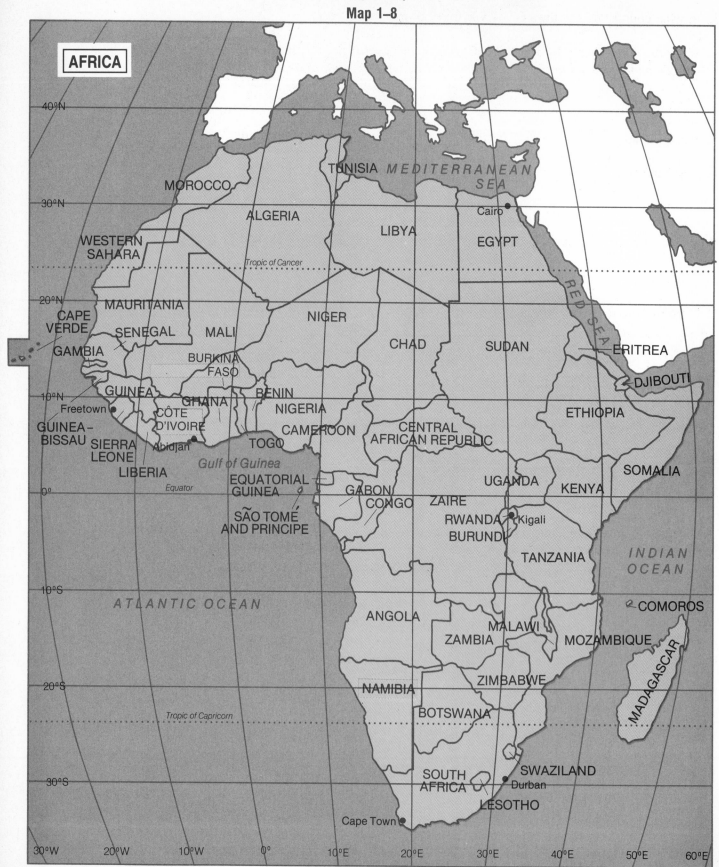

AFRICA

1. What body of water is located at 0° latitude, 0° longitude?   _____the Gulf of Guinea_____

2. What city is located near 30°S latitude, 30°E longitude?   _____Durban_____

3. In what country do the lines of 10°N latitude, 0° longitude cross?   _____Ghana_____

4. Write the approximate location of Cairo, Egypt, using latitude and longitude. (Remember that the lines of latitude and longitude on this map are spaced 10° apart. Give locations to the nearest degree.)

   _____30°N latitude, 32°E longitude_____

5. Write the approximate location of Kigali, Rwanda, using latitude and longitude.   _____2°S latitude, 30°E longitude_____

6. Write the approximate location of Cape Town, South Africa, using latitude and longitude.

   _____34°S latitude, 18°E longitude_____

7. Write the approximate location of Abidjan, Côte d'Ivoire, using latitude and longitude.

   _____5°N latitude, 5°W longitude_____

8. Write the approximate location of Freetown, Sierra Leone, using latitude and longitude.

   _____9°N latitude, 12°W longitude_____

C   Use a map of your state that includes lines of latitude and longitude to complete the following activities.

1. Write the approximate location of your city using latitude and longitude.   _____Answers will vary._____

2. Write the approximate location of your state capital using latitude and longitude.

   _____Answers will vary._____

3. Select a popular recreation area in your state, such as a National Park, National Forest, or large lake. Write the name of this recreation area and its approximate location using latitude and longitude.

   _____Answers will vary._____

Lesson 6   Locating Continents and Oceans

OBJECTIVE

Locate major landmasses and bodies of water in many parts of the world

The surface of the earth is covered with land and water. The land is divided into seven continents: North America, South America, Europe, Asia, Africa, Australia, and Antarctica. The continents are divided into over 190 nations. The water is divided into four oceans and a number of seas. The four oceans are the Atlantic, Pacific, Indian, and Arctic oceans.

Before you proceed with this lesson, you should study the locations of the continents, major countries, and oceans on a world map.

Using Your Skills

A   The continents and oceans are labeled with letters on Map 1–9. Write the name of each continent or ocean beside the correct letter below.

1. A   Antarctica

2. B   Asia

3. C   North America

4. D   Australia

5. E   Pacific Ocean

6. F   Atlantic Ocean

7. G   Indian Ocean

8. H   Africa

9. I   Arctic Ocean

10. J   South America

11. K   Pacific Ocean

12. L   Europe

13. M   Atlantic Ocean

B   Fill in the blanks to complete the following sentences correctly.

1. South America is bordered by the _____Pacific_____ Ocean on the west and the

_____Atlantic_____ Ocean on the east.

2. Africa is _____south_____ of Europe.

3. The most direct route from Australia to Africa is across the _____Indian_____ Ocean.

24

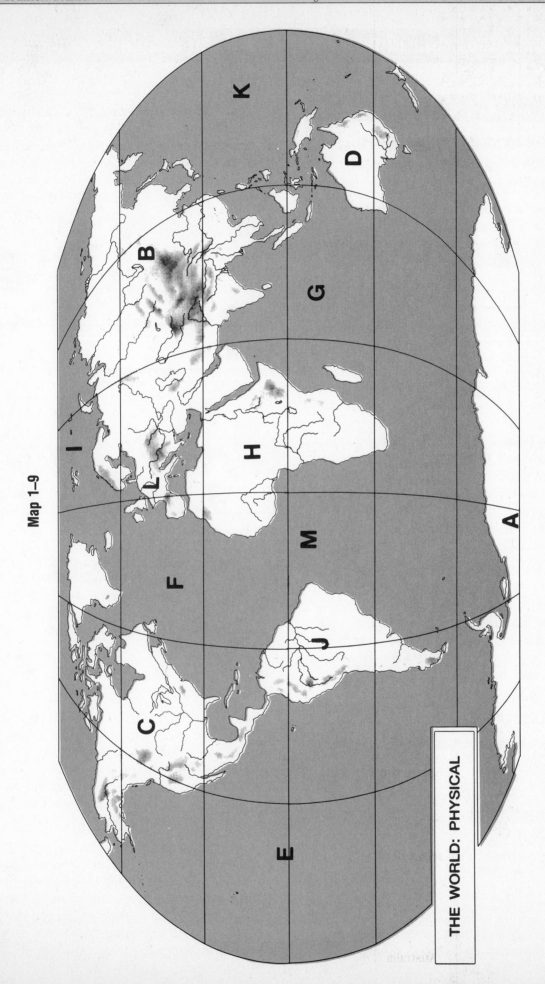

Map 1–9

THE WORLD: PHYSICAL

OBJECTIVE

Use map legends to interpret symbols commonly used on maps

TERMS TO KNOW

key (kee)—part of a map that tells the meaning of symbols

legend (LEJ·uhnd)—part of a map that tells the meaning of symbols

symbol (SIM·buhl)—drawing used on a map

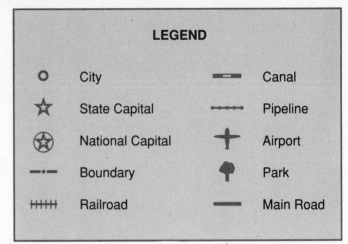

Example 1–7

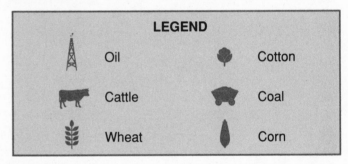

Example 1–8

Maps can be used to show many different kinds of information. One of the main uses of maps is to show the locations of towns and cities. However, maps can also show where roads, parks, motels, and many other things are located. Maps can show where cotton is grown, cattle are raised, and kinds of wild animals are found. The possibilities are almost endless.

Often one map shows more than one kind of information. This means that some way must be used to help the person reading the map understand several kinds of information. For example, if a map shows the locations of towns, roads, and parks, there must be some way for the person reading the map to know which is which.

Using Map Legends

Maps use **symbols** to help the reader tell kinds of information apart. The meaning of each symbol is explained in a special part of the map called a **legend,** or **key.** Each symbol used on the map is shown, along with an explanation of what the symbol means.

Look at Example 1–7. What symbol is used to show the location of a city? What symbol is used to show the location of a city which is a state capital? What symbol is used to show the location of a park?

There can be many different symbols. Some maps show the kinds of products a state or country

produces. Often these maps use picture symbols to show where products are produced. For example, a small picture of an oil derrick ♦ may be used to show where oil is found. A picture of a sheep may be used to show where sheep are raised.

Look at Example 1–8. What symbol shows where wheat is grown? What is grown where you see the symbol 🌳 ?

Sometimes maps use colors or areas of shading as symbols. This is often used when the feature being shown covers a wide area. Here are some examples of shading: ▨ ▦ When you are reading a map which uses shading, you must be very careful to read the map correctly. It is easy to get shading like the following mixed up. ▨ ▨

If you have trouble telling such patterns apart, try this. Look at the part of the map you wish to read. Then look at the legend and pick the pattern you think is correct. Cover the others with your fingers or a piece of paper. Look at the legend and then at the map. This should help you decide if you have picked the correct pattern.

Using Your Skills

A Use map legends 1–7 and 1–8 to match each symbol below with its meaning.

___c___ 1. ⬛   a. wheat

___e___ 2. ⌗⌗⌗⌗   b. airport

___f___ 3. ☆   c. corn

___a___ 4. 🌾   d. oil

___b___ 5. ✈   e. railroad

___d___ 6. 🗼   f. state capital

B Use Map 1–10 and its legend to answer the questions which follow.

1. What covers most of the Amazon River basin? ___tropical hardwood forest___

2. In how many places in the Amazon River basin is gold found? ___four___

3. In what part of the Amazon River basin is coal found? ___in the northwestern part___

4. About how many miles apart are the deposits of tin in the Amazon River basin? ___1,000___

5. How many deposits of iron ore are shown in the Amazon River basin? ___two___

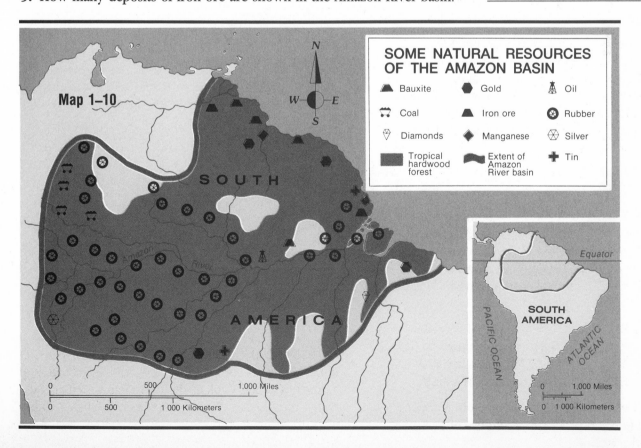

Map 1–10

SOME NATURAL RESOURCES OF THE AMAZON BASIN

- Bauxite
- Coal
- Diamonds
- Tropical hardwood forest
- Gold
- Iron ore
- Manganese
- Extent of Amazon River basin
- Oil
- Rubber
- Silver
- Tin

OBJECTIVE

Use a road map to select routes and estimate distance and travel time, and plan a family trip

A new thing from American carmakers is a car that doesn't need road maps. A special computer in the car "talks" to a satellite orbiting the earth. The computer and the satellite keep track of where the car is every minute. This information is displayed on a screen in the car. As you drive the car, its location is shown on a map on the screen.

Using Road Maps

Even with one of these cars, however, we will still need road maps. Road maps are useful for planning what route to take on a trip, finding the location of places that are new to us, and estimating how long it will take to drive to a place. Road maps can also give information about things to see and do.

Reading a road map requires many of the skills you have practiced in earlier lessons in this book. You need to know how to find direction and distance. Road maps have an index that uses a grid. Symbols are used on road maps to tell you such things as how large towns are, where you can stop to rest, and even points of interest along the way.

You will often use the index first when you read a road map. Find the name of the place you want to go to in the index. The index will tell you the cell in the grid where the place is located.

Map 1–11 Road Map of Part of New York State

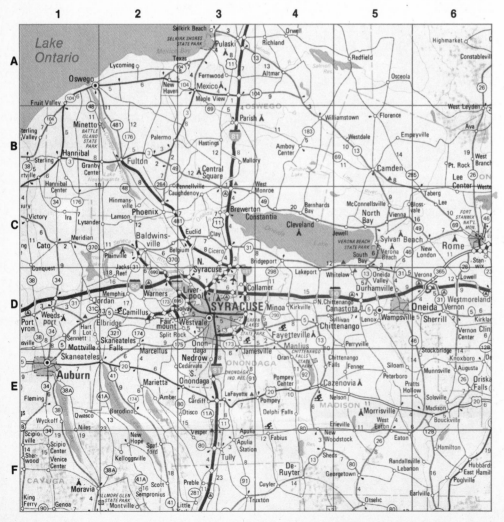

LEGEND

⛷	Ski Area
▲	Rest Area Without Restrooms
Ⓐ	Rest Area With Restrooms
𝝠	Campground
✈	Airport
	Interstate Highway
	U.S. Highway
	State Highway
	Distance Between Mileage Markers

Using Your Skills

A   **Use the legend of Map 1–11 to answer these questions.**

1. What kind of highways are the ones numbered 81 and 90 which cross near the center of the map?

interstate highways

2. What can you expect to find at a place marked with this symbol ____ ?   a campground

3. What do the numbers between two marks like these ____ tell you?   how many miles it is between the two marks

4. What is the difference in meaning between these two symbols ____ ?   The triangle with the circle around it means a rest stop with rest rooms. The triangle by itself means a rest stop without rest rooms.

5. What could you expect to do at a place marked with this symbol ____ ?   go skiing

B   **Use Map 1–11, its legend, and the map index to answer these questions.**

1. In which cell in the grid are each of the following located?

a. Parish _____ B-3   **b.** Cleveland _____ C-4

c. Texas _____ A-3   **d.** Chittenango Falls State Park _____ E-4

2. What could you expect to do near Fabius (F-4) in January?   go skiing

3. Plan a trip from Mexico to Weedsport using interstate highways as much as possible. How far will you travel?

54 miles

4. Suppose that you average about 50 miles per hour on interstate highways. About how long will it take you to travel from the farthest point north shown on the map to the farthest point south?

about $1\frac{1}{2}$ hours

C   **On your own paper, plan a vacation trip. Use a highway map of your own state. Start at your home town and trace out a three-day trip of places you would like to visit. What points of interest would you see? What towns would you go through? How many miles would you travel over which highways? Think about where you might stay overnight and good places to stop for food, too.**

Answers will vary.

OBJECTIVE

Analyze how basic earth-sun relationships affect everyday life

TERMS TO KNOW

axis (AX·is)—imaginary lines drawn from the North Pole to the South Pole

Northern Hemisphere (NOR·thurn HIM-uhs·fehr)—part of the earth north of the equator

Southern Hemisphere (SUH·thern HIM-uhs·fehr)—part of the earth south of the equator

Our seasons are caused by the relationship between earth and the sun. The sun plays a very important part in our lives. Life on earth is supported by energy from the sun. Our language shows how important the sun is to our everyday lives. We have sun roofs, sunglasses, and sunsuits. We need the Vitamin D that the sun provides, to keep our bodies strong.

What Causes Seasons?

If all parts of the earth received the same amount of sunshine year around, there would be no seasons. However, this is not the case. The difference in the amount of sunshine received is due to two things: the movement of the earth around the sun, and the tilt of the earth on its **axis.** The earth's axis is an imaginary line drawn from the North Pole to the South Pole through the center of the earth.

Seasons in the Northern Hemisphere

Look at Example 1–9 of the earth's movement around the sun. This diagram shows the seasons for the **Northern Hemisphere,** the part of the earth north of the equator. This is where the United States is.

What is your favorite season of the year? Do you like summer more than winter?

Do you know *why* we have seasons? Do you know that you can go from one season to another just by changing your location?

Example 1–9

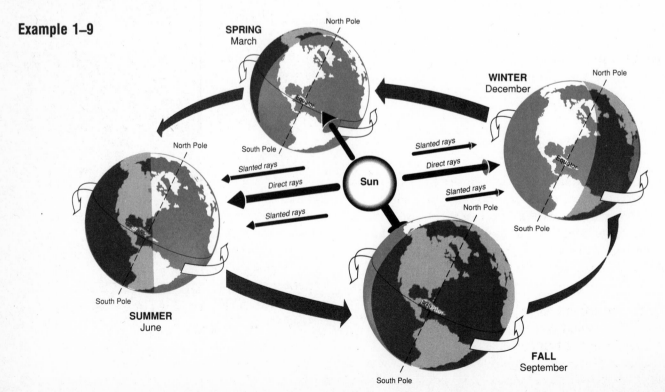

The earth spins on its axis once in 24 hours. This is the length of our day. The earth also travels around the sun. This trip takes a little over 365 days—one year. Notice that the earth's axis is tilted. As the earth travels around the sun, its axis always points in the same direction. This means that sometimes the North Pole is pointed toward the sun. At other times the North Pole is pointed away from the sun.

Find the position of the earth marked "summer." Place a pencil on the drawing so that its ends line up with the North and South poles. Notice that the upper end of the pencil is tilted toward the sun in the drawing. The North Pole points toward the sun when we are having our summer.

Now find the position of the earth marked "winter." *Without changing the angle at which your pencil is tilted,* move the pencil to the earth's winter position. Notice that now the upper end of the pencil is tilted away from the sun. The North Pole points away from the sun when we are having our winter.

Seasons in the Southern Hemisphere

When it is summer in the Northern Hemisphere, it is winter south of the equator, in the **Southern Hemisphere.** For example, if you are baking in the heat of an Arizona summer, you can fly to somewhere south of the equator and enjoy winter weather.

Whether we are having summer or winter is due to how much heat our location on the earth is getting from the sun. When the northern half of the earth is pointing toward the sun, it gets more heat for two reasons. One reason is that there are more hours of daylight. The other reason is that the sun's rays fall more directly on the northern hemisphere. These direct rays are hotter than the sun's rays in winter.

How the Sun Affects Seasons

You can use two pieces of paper, or two paper plates, to show why the direct rays of the sun are hotter. Cut a circle about 1½ inches across in the first piece of paper. Hold this piece of paper up to the sun or a strong lamp. Hold the second piece of paper about 3 inches behind it so that light comes through the hole. This represents the light of the sun falling on the earth.

Now tilt the second piece of paper so that first it points straight up and down, and then so that its top points away from the first piece, and then toward it. What happens to the spot of light falling on the paper? It changes shape. When the top of the second paper is

tilted toward or away from the first piece, the spot of light gets bigger.

The smaller the spot of light, the hotter it is. In summer the sun is almost directly overhead. Its rays fall on a smaller spot of earth than in winter. That part of the earth gets more heat. In winter the sun is lower in the sky. Its rays are slanted and fall on a larger spot of earth than in summer. That part of the earth gets less heat.

What happens if the sun is always almost directly overhead? There will be no seasons. The weather will always be warm. This happens near the equator.

Using Your Skills

A Use Example 1–9 of the earth's movement around the sun to help you label Map 1–12 correctly.

Map 1-12 Seasons in the Northern Hemisphere

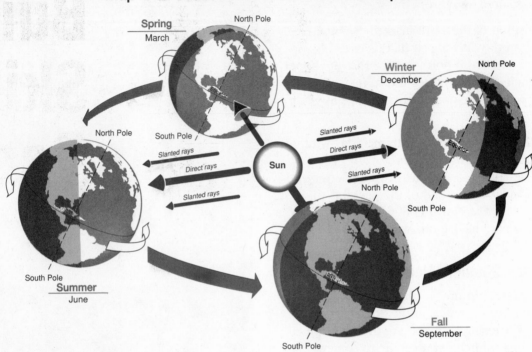

B Seasons in the Southern Hemisphere are opposite those in the Northern Hemisphere. Label Map 1–13 to show the seasons in the Southern Hemisphere.

Map 1-13 Seasons in the Southern Hemisphere

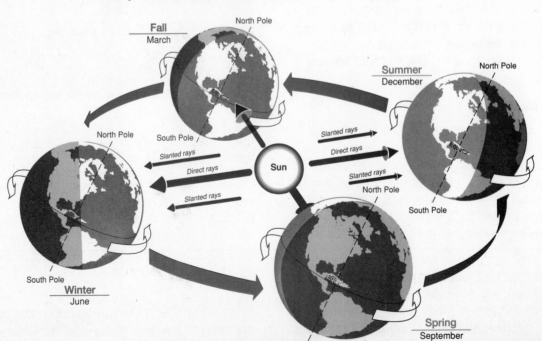

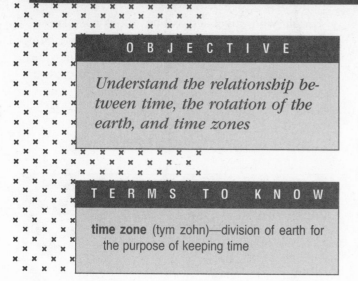

Would you like to be able to fly through space at over 1,000 miles per hour? Well, you are—right this minute. The earth spins on its axis at about 1,000 miles per hour at the equator, carrying you with it. Each hour your spot on the earth travels 15 degrees of longitude toward the east.

You may think you do not notice any sign of your speedy trip, but you do. Every day you see the sun march across the sky. The sun is not actually moving, of course. The earth is turning from west to east. That is why the sun comes up in the east and sets in the west.

It takes the earth 24 hours to turn on its axis once. Imagine that the sun has just come up. In one hour the sun will be higher in the sky. With each hour that passes, the sun will rise higher, until noon. Then it will become lower, until finally it sinks out of sight.

Imagine that you have a friend who lives 1,000 miles west of you. When the sun has been up for one hour where you live, it will just be coming up where your friend lives. You have another friend who lives 1,000 miles east of you. When the sun has been up for one hour where you live, it will have been up for two hours at your friend's house to the east.

Using the Sun to Tell Time

People have used the sun to tell time for many years. How high the sun is in the sky can tell us how long it has been since sunrise, and how long it is until sunset. When the sun is at its highest point in the sky, it is noon. Remember your two friends to the east and the west of you. When it is noon where you are, it is an hour past noon where your friend to the east lives. It is an hour before noon at your friend's house to the west.

Look at the bar on the bottom of the page. It shows the 24 hours in a day. The sun is at 12 noon. Each division of the bar stands for the distance the earth turns in one hour. As you move east on the bar, times become later in the day. As you move west, times become earlier. For example, one division east of 12 noon, the time is 1:00 P.M. One division west, the time is 11:00 A.M. Fill in the blanks on the bar to show the correct times. Cut the bar out and tape it around a tennis ball. This shows you how time changes as you go around the world. Notice that when it is 12 noon on one side of the world, it is 12 midnight on the other.

Time Zones

The earth is divided into 24 parts for keeping time, just like the bar above on the edge of the page. We call each division of the earth a **time zone.** Every place on the earth within a time zone has the same time as every other place in that zone.

Before we had time zones, every town kept its own time. Because of the earth's rotation, noon came at different times for towns even forty or fifty miles east or west of each other. As long as travel was slow, this was not a problem. But with the coming of railroads, the differing times became a big problem. Trying to tell people when trains would arrive and leave was almost impossible when the clocks in every town were set at a different time.

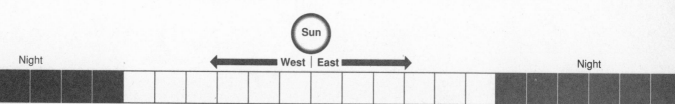

Night   West | East   Night

Time zones were set up to solve this problem. Time zones are about 1,000 miles across from east to west at the equator. Time zones become narrower as you move toward the poles. Only four time zones are needed to cover the entire continental United States. (Alaska and Hawaii are in other time zones because they are farther west.) These four times zones are called the Eastern Time Zone, the Central Time Zone, the Mountain Time Zone, and the Pacific Time Zone. In some cases the time zones follow the boundaries of states or nations rather than lines of longitude. Find these zones on Map 1–14.

People who travel across time zones must keep track of time. Whenever you cross a time zone going east, the time becomes one hour *later*. You must set your watch ahead one hour. Whenever you cross a time zone going west, the time becomes one hour *earlier*. You must set your watch back one hour.

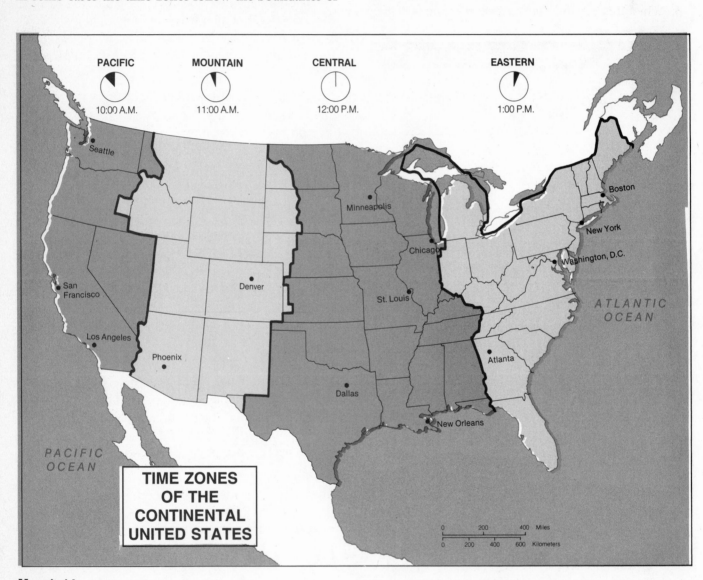

Map 1–14

Using Your Skills

A   **Use Map 1–14 of the United States to answer the questions.**

1. Which time zone is farthest east?   the Eastern Time Zone

2. Which time zone is farthest west?   the Pacific Time Zone

3. When it is 12 noon in Dallas, what time is it in New York City?   1:00 P.M.

4. When it is 4:00 P.M. in Denver, what time is it in San Francisco?   3:00 P.M.

5. Suppose that you live in Atlanta. Your grandparents in Seattle want to call you on your birthday. They go to sleep at 10:00 P.M. What is the latest Atlanta time you can expect to hear from them?

   7:00 P.M.

6. Suppose you live in St. Louis. You have a computer made by a company near Los Angeles. You want to call them about a problem you are having with your computer. They go to work at 9:00 A.M. What is the earliest

   St. Louis time you can call them?   11:00 A.M.

7. Imagine that you are flying from Boston to San Francisco. You leave Boston at 8:00 A.M. What time is it in San

   Francisco?   5:00 A.M.

8. The plane trip from Boston to San Francisco takes six hours. You leave Boston at 8:00 A.M. What time will it

   be in Boston when you land?   2:00 P.M.

   What time will it be in San Francisco when you land? Why?   11:00 A.M. You gain one hour for each time zone you

   pass through going west.

9. You have to fly from San Francisco to Chicago. You leave San Francisco at 5:30 P.M. What time is it in

   Chicago?   8:30 P.M.

10. The plane trip from San Francisco to Chicago lasts 4 hours. You leave San Francisco at 5:30 P.M. What time

    will it be in Chicago when you land? Why?   12:30 A.M. Whenever you cross a time zone going east, the time becomes

    one hour later.

Lesson 11   Comparing Types of Maps

Suppose you are planning a trip to another country. What would you like to know about that country before you leave home? What language is spoken there?

Where are interesting places to visit located? Should you take clothes for warm weather, or cold?

These are just a few of the things you can learn from maps. In fact, there are almost as many kinds of maps as there are kinds of information to show on them. Let's look at some of the different kinds of maps and what they can tell us about our world.

Using Physical Maps

A **physical map** shows how the land looks. Mountains, rivers, plains, and lakes—the physical features of the land—are shown on a physical map. Sometimes a physical map shows the height of the land above sea level. This kind of physical map is called a **relief map.**

Look at Map 1–15 of the United States. Find the Rocky Mountains. What physical feature lies just east of the Rocky Mountains? Into what river do the Ohio and Missouri rivers flow? Into what body of water does the Mississippi River flow? What ocean lies east of the United States? These are all kinds of information you can find on physical maps.

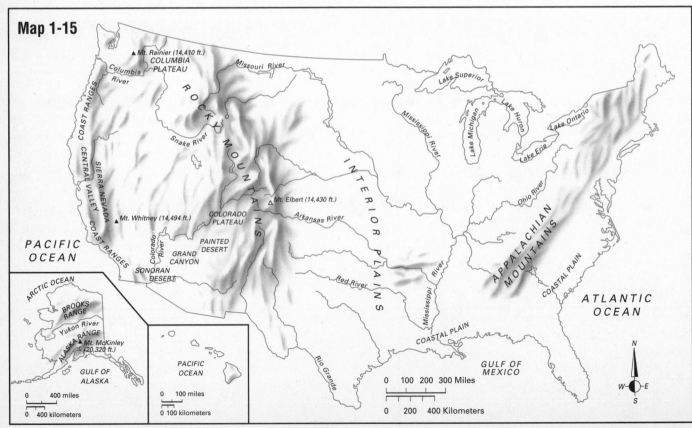

Map 1-15

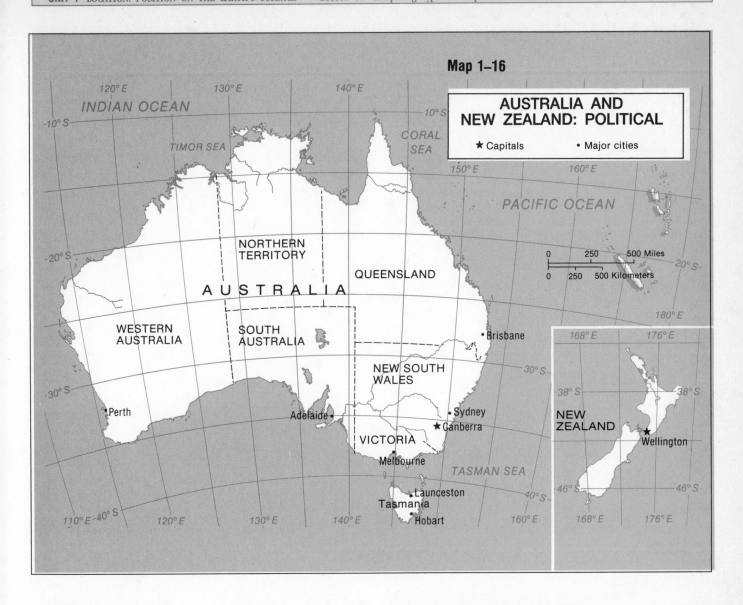

Map 1–16

AUSTRALIA AND NEW ZEALAND: POLITICAL

★ Capitals   • Major cities

Using Political Maps

A **political map** shows how humans have divided the surface of the earth into countries, states, and other pieces. Often, a political map will show some physical features, such as lakes and rivers, because these are sometimes used as boundaries. A political map will show where the boundaries between countries, states, or counties are. It may also show the locations of cities.

Look at Map 1–16 of Australia. Notice the dashed lines on the map. These show the boundaries between states. The letters in SMALL CAPITALS are the names of the states. Queensland is the name of one state. Can you name the others? What is the capital of Australia? What tells you it is the capital? What are some other cities shown on the map?

Using Special-Purpose Maps

There are many kinds of maps that serve special purposes. Some of the kinds of **special-purpose maps** are rainfall maps, product maps, and population density maps. These maps give one particular kind of information. You will learn to read these and other kinds of special-purpose maps later in this book.

Many times maps will be a combination of physical, political, and special-purpose. For example, a special-purpose map which shows what products are produced in the United States will usually have state boundaries shown. What kind of map shows boundaries? A product map may also show major rivers. What kind of map shows rivers?

When you read a map, what do you think should be the first thing you should look at? In order to know what the map is about, you must look at the map's title. The title may be at the top or bottom of the map, or it may be in a box with the legend.

Using Your Skills

A   **Match each term with its meaning. Draw a line from each term to its definition.**

1. political map

2. physical map

3. special-purpose map

    a. a kind of map which shows features of the land

    b. a kind of map which gives one particular kind of information

    c. a kind of map which shows how humans have divided the earth among themselves

B   **Use Map 1–17 to answer the questions.**

1. What is the title of this map? _____ Ethiopia _____

2. What part of the map tells you what the symbols on the map mean? _____ the legend _____

3. What do the solid black lines on the map stand for? _____ national boundaries _____

4. What does ≈≈≈≈≈≈≈≈ on the map stand for? _____ mountains _____

5. What does 🚢 on the map stand for? _____ seaports _____

6. Would you call this a physical map, a political map, a special-purpose map, or a combination of all three? Why?

It is a combination of all three types. It shows political boundaries and cities. It shows physical features such as mountains and rivers. It

also shows special features such as roads, railroads, airports, and seaports.

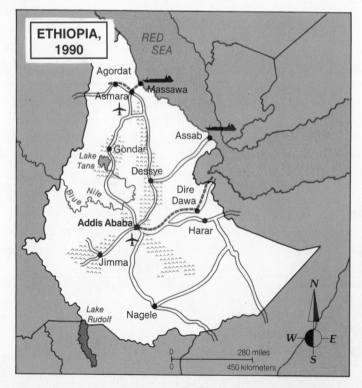

Map 1–17

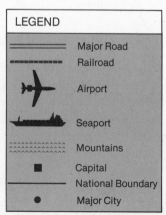

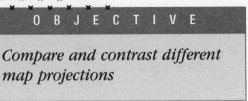

*Compare and contrast different
map projections*

T E R M S   T O   K N O W

conic projection (KAHN·ik proh·JEK·
shuhn)—map projection used for showing
small areas midway between the equator
and the poles

Gall-Peters projection (gawl PEE·ters
proh·JEK·shuhn)—map projection that
shows the sizes of landmasses correctly

map projection (map proh·JEK·shuhn)—a
way of showing the earth on a piece of
paper

Mercator projection (murh·KAYT·er
proh·JEK·shuhn)—map projection that
shows true directions and land shapes but
exaggerates sizes of landmasses

H ave you ever heard the saying, "You can't have
your cake and eat it, too"? It's a way of saying
that when we get one thing, often we must give up
something else.

Choosing a map sometimes means giving up one
thing in order to get another. Maps show four things:
<u>direction</u>, <u>distance</u>, <u>shape</u>, and <u>size</u>. Only a globe can
show all four with accuracy at the same time. Maps,

however, cannot. A map may show direction well, but
the shapes of landmasses may be quite inaccurate. Or,
if shapes are shown correctly, distances may not be.

You might think that the way to get around this
problem would just be to use globes all the time.
However, think how hard it would be to get a globe
in your pocket or inside the covers of a book.

Understanding Map Projections

There are many different kinds of **map
projections.** A map projection is a way of showing
the rounded earth on a flat piece of paper. Where
does the word *projection* come from? Imagine a clear
globe with latitude and longitude lines and the
outlines of the landmasses on it. Suppose there was a
light bulb inside the globe. If you wrapped a piece of
paper around the globe and turned on the light bulb,
the outlines of the grid and landmasses would be
projected onto the paper.

Mercator Projection

Look at Example 1–10 of a **Mercator projection.**
This type of projection shows how the earth would
look if a piece of paper were wrapped to form a tube
around the globe. You will recall that lines of latitude
are the same distance apart on a globe. But look at
what happens to lines of latitude on a Mercator
projection. The lines get farther apart as you go away
from the equator. This means that distances are not
true. It also means that the sizes of landmasses near
the North and South poles are greatly exaggerated in
size. For example, South America is actually almost
nine times as large as Greenland. On a Mercator map,
however, Greenland looks bigger than South America.

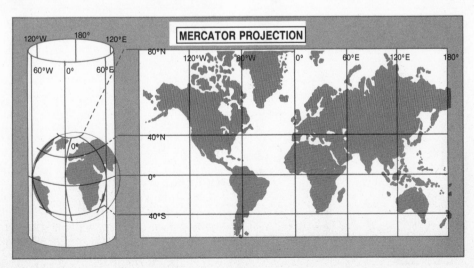

Example 1–10

Gall-Peters Projection

A **Gall-Peters projection** looks similar to the Mercator projection. Latitude and longitude lines on both are shown as straight lines that cross at right angles. Unlike a Mercator projection, the sizes of landmasses are accurate on a Gall-Peters projection. Shapes and distances are not correct, however.

Example 1–11

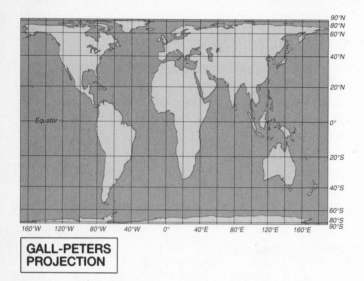

GALL-PETERS PROJECTION

Conic Projections

Often you will see maps on which the longitude lines are straight and get closer together toward the north, or top of the map. Latitude lines are curved on this kind of map. It is called a **conic projection.** It comes from the idea of placing a cone over part of a globe.

A conic projection is good for showing small areas midway between the equator and the poles. Size, distance, and direction are fairly accurate.

Example 1–12

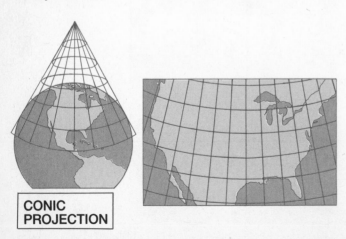

CONIC PROJECTION

There are a number of other kinds of projections which show the sizes of landmasses fairly accurately. Look at Examples 1–13 through 1–15. Notice that each has a particular shape.

Remember that no map can show direction, distance, shape, and size at the same time as accurately as a globe. Every kind of map has a special use, but none is perfect. When you look at a map, keep in mind that the sizes of landmasses may not be correct, or distances or directions may not be true. Be careful not to make judgments about the world based only on maps.

Example 1–13

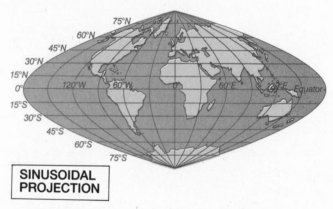

SINUSOIDAL PROJECTION

Example 1–14

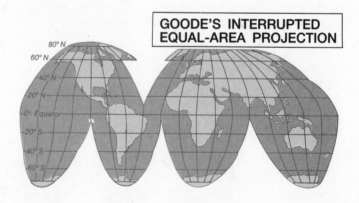

GOODE'S INTERRUPTED EQUAL-AREA PROJECTION

Robinson Projection

A **Robinson projection,** Example 1–15, is the most widely used map projection of the world. It shows only minor distortions in true size, distance, and shape of landmasses.

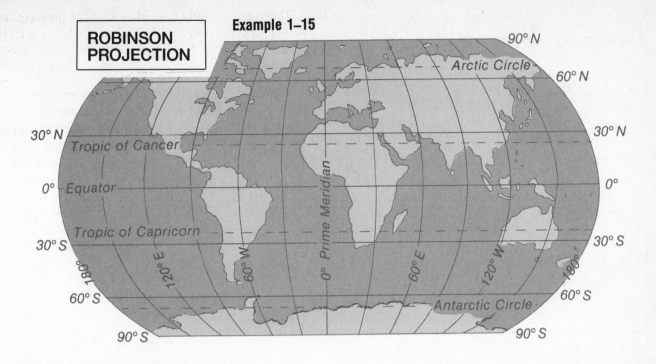

ROBINSON PROJECTION

Example 1–15

Using Your Skills

A Decide whether each statement is true or false. Write *T* if the statement is true. Write *F* if the statement is false.

_____ T _____ **1.** Only globes can show true distance, direction, size, and shape of landmasses all at the same time.

_____ F _____ **2.** No map can show true distance and direction.

_____ F _____ **3.** All maps show the surface of the earth in the same way.

_____ F _____ **4.** All maps carry a legend which tells you what kind of projection was used to make the map.

_____ T _____ **5.** A map can show that a landmass is much larger than it really is.

B Fill in the blanks to correctly complete the following sentences.

1. On a Mercator projection, the sizes of landmasses near the North and South poles are greatly exaggerated in

_____ size _____

2. On a Gall-Peters projection, the sizes of landmasses are _____ accurate _____, but shape and distances

are _____ not _____   _____ correct _____

3. A conic projection is good for showing _____ small _____ areas midway between the

_____ equator _____ and the poles.

Lesson 13   Historical Change in Importance of Location

OBJECTIVE

Be aware that the significance and importance of locations change as cultures change their interactions with each other and with the physical environment

The soldiers creeping to the small Pennsylvania town on June 30, 1863, had no wish to make the town famous. All they wanted was to find shoes, for they had none. But five days later the name *Gettysburg* had become one which would be written in history. One of the greatest battles of the American Civil War began there simply because soldiers of the two sides bumped into each other almost by accident. Today, Gettysburg is the site of a national military park which attracts thousands of visitors a year.

Why did the importance of Gettysburg's location change? The importance of Gettysburg's location changed because of events that took place there. The town is far more important today than it would be if the battle had been fought somewhere else.

Humans interacting with each other can change the importance of a location. The location can become more or less important. The change can take place quickly, as in the case of Gettysburg, or slowly.

The Story of Timbuktu

The African city of Timbuktu is a good example of a location whose importance changed slowly. At first a small village, Timbuktu grew into one of the foremost cities of its time. Today, it is once again a village. What happened, and why, is one of the most interesting stories in geography.

Timbuktu is located in the western part of Africa South of the Sahara, on a bend in the Niger River, at about 17°N latitude, 3°W longitude. Can you see anything about this location that would explain why Timbuktu became a great city? Other than Timbuktu's location on a river, there is very little to go on.

Trade's Impact on Timbuktu

Look at Map 1–18 showing trade routes in West Africa about the year 1000 A.D.—almost 1,000 years

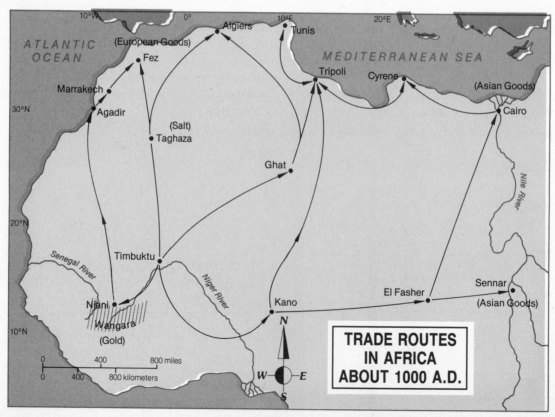

Map 1–18

TRADE ROUTES
IN AFRICA
ABOUT 1000 A.D.

ago. Find the area marked Wangara. Gold was mined here. Now find the village of Taghaza to the north. Salt was mined here. Find Timbuktu, which is between Wangara and Taghaza.

You can see that Timbuktu is located about halfway between gold mines to the south and salt mines to the north. The people who had gold needed salt, and the people who had salt wanted gold.

By now you may suspect that trade, not just location, made Timbuktu a great city. However, it was Timbuktu's location that made it be chosen as the place where the trade took place.

The gold and salt mines operated for hundreds of years. The gold was carried north to the Niger River. There traders from the south met traders from the north. The gold was traded for many kinds of goods from Europe—cloth, swords, beads, horses, and foods—but especially for salt. Salt was a necessary food. It was very important to these people in a hot climate, who lost much salt through sweat each day.

The people of the area around Timbuktu needed salt so badly that sometimes they would trade a weight of gold for the same weight of salt. The salt that was so precious in Timbuktu was so common in Taghaza that houses were built of blocks of it.

The king of the area around Timbuktu claimed much of the gold from Wangara for himself. He also taxed all the gold and salt that were brought in to be traded. The king and the traders became very rich.

Timbuktu Becomes a Center of Learning

Timbuktu became not only a center of trade, but also a center of learning. The rulers of Timbuktu became Muslims, followers of the religion of Islam. Muslims believe in education, because they believe that Muslims should be able to read the Quran, the book of Muslim teachings. Many Muslim traders came to live in Timbuktu. They brought their love of learning with them. The city became known for its teachers and libraries. One visitor wrote that "Here are a great store of doctors, judges, priests, and other learned men, that are bountifully maintained at the king's cost and charges. And hither are brought [many] manuscripts of written books . . . which are sold for more money than any other merchandise. . . ."

Timbuktu was a great city for hundreds of years. However, the riches of the area attracted many invaders. Shortly before the year 1600, an army from Morocco attacked. Over a period of many years, the area around Timbuktu was ruled by many different countries. Trade was broken up by wars. The gold mines of Wangara ran out. Timbuktu once again became a poor village.

Timbuktu Today

Timbuktu is no longer a center of world trade and learning. It is still located at about 17°N latitude, 3°W longitude. Its location has not changed, but the *importance* of its location has changed.

The salt mines at Taghaza still produce salt for the people of Western Africa. People come on camels and in jeeps to buy the salt. But even the importance of the location of Taghaza has changed. No longer is it a stop on a trade route linking Europe with Timbuktu. No longer does gold from Timbuktu flow through Taghaza on its way to the coast to buy goods from Europe.

Using Your Skills

Answer these questions about the reading and map.

1. What was important about Timbuktu's location? It was located midway between salt mines and gold mines. People

traded gold for the salt and other goods. They met at Timbuktu to trade.

2. How can people change the importance of a location?   by interacting there
_____

3. How was the importance of Taghaza's location changed? How has it stayed the same?   The importance of

Taghaza's location is different because it is no longer part of a trade route from Timbuktu to the north coast of Africa. It is the same

because people still come there to buy salt.
_____

_____

4. How did trade change Timbuktu?   Timbuktu became a great city. Many people came to live there. Many goods were

traded there. Many people who lived there became rich. The city became a center of learning as well as a center of trade.
_____

_____

5. About how many miles did people travel from the north coast of Africa to trade at Timbuktu?   about

2,000 miles
_____

6. About how many miles did traders from Asia travel across Africa to trade at Timbuktu?   about 2,500 miles
_____

7. Almost all the area traders had to cross to trade at Timbuktu was covered by desert. What does this tell

you about how important the trade was to them?   The trade was very important to them.
_____

8. Is your town an important center of trade?   Answers will vary.
_____

9. For what is your town or city known?   Answers will vary.
_____

_____

_____

10. What events or historic changes have affected the importance of your town or city?   Answers will vary.
_____

_____

_____

OBJECTIVE

Learn to use maps to know the correct location of physical and human features of the United States

Using Your Skills

There are 25 errors in Map 1–19 of the United States. Some states and bodies of water are mislabeled. Compare the map to a correct map and underline each label you think is in error on the map. States with no labels are not to be counted as an error. Missing states are also considered an error.

Map 1–19

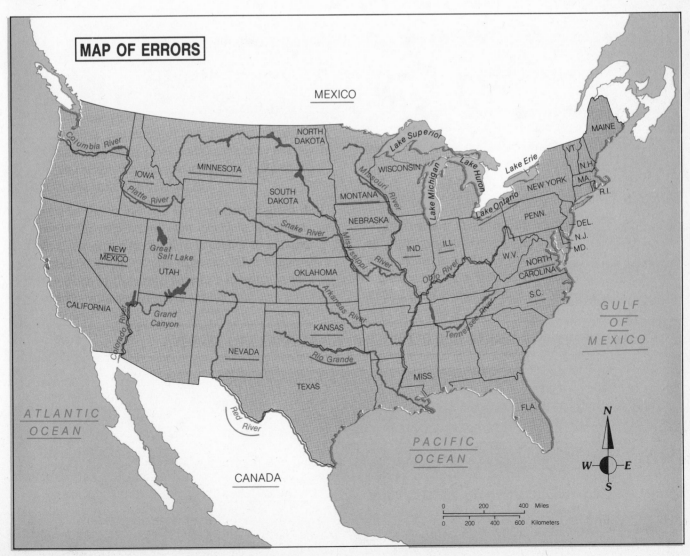

MAP OF ERRORS

MEXICO

Columbia River

NORTH DAKOTA

Lake Superior

MAINE

WISCONSIN

Lake Huron

Lake Erie

VT

N.H.

IOWA

MINNESOTA

Missouri River

Lake Michigan

NEW YORK

MA.

R.I.

Platte River

SOUTH DAKOTA

MONTANA

Lake Ontario

PENN.

NEBRASKA

DEL.

Snake River

N.J.

NEW MEXICO

Great Salt Lake

IND.

ILL.

MD.

W.V.

NORTH CAROLINA

Mississippi River

Ohio River

UTAH

OKLAHOMA

CALIFORNIA

Colorado River

Grand Canyon

Arkansas River

S.C.

GULF OF MEXICO

KANSAS

Tennessee River

NEVADA

Rio Grande

MISS.

TEXAS

FLA.

ATLANTIC OCEAN

Red River

PACIFIC OCEAN

N

CANADA

W   E

S

0   200   400   Miles

0   200   400   600   Kilometers

45

U N I T 1 R E V I E W

A   **Underline the term in parentheses in each sentence which will complete the statement correctly.**

1. Map makers use a (<u>compass rose</u>, legend) to show direction on a map.

2. Maps usually have a (grid, <u>scale</u>) to show what distance on the earth is represented by a certain distance on the map.

3. The location of a place on the earth as compared to some other place is called (absolute location, <u>relative location</u>).

4. Depending on other people is called (<u>interdependence</u>, socialism).

5. Distance north and south of the equator is measured in degrees of (<u>latitude</u>, longitude).

6. Distance east or west of the prime meridian is measured in degrees of (latitude, <u>longitude</u>).

7. Another name for the prime meridian is (0° latitude, <u>0° longitude</u>).

8. The legend of a map tells what the (<u>symbols</u>, cells) on the map mean.

9. When it is winter in the Northern Hemisphere, it is (spring, <u>summer</u>) in the Southern Hemisphere.

10. When it is noon where you are, it is 2:00 P.M. two time zones to the (<u>east</u>, west).

11. A map which shows how the land looks is called a (<u>physical</u>, political) map.

12. A product map is an example of a (relief, <u>special-purpose</u>) map.

13. Direction, distance, shape, and size can all be shown correctly at the same time only on a (<u>globe</u>, map).

B   **Follow the instructions below as you label Map 1–20 on the next page.**

1. Find the prime meridian. Label it 0° longitude.

2. Find the equator. Label it 0° latitude.

3. Label the continents.

4. Label the Atlantic, Pacific, and Indian oceans.

C   **Use Map 1–20 on the next page to answer these questions.**

1. Name the two continents located entirely south of the equator.   <u>Australia, Antarctica</u>

2. Name the continents crossed by the prime meridian.   <u>Europe, Africa</u>

3. Much of Europe and Asia are at the same latitude as   <u>North America</u>

Map 1–20

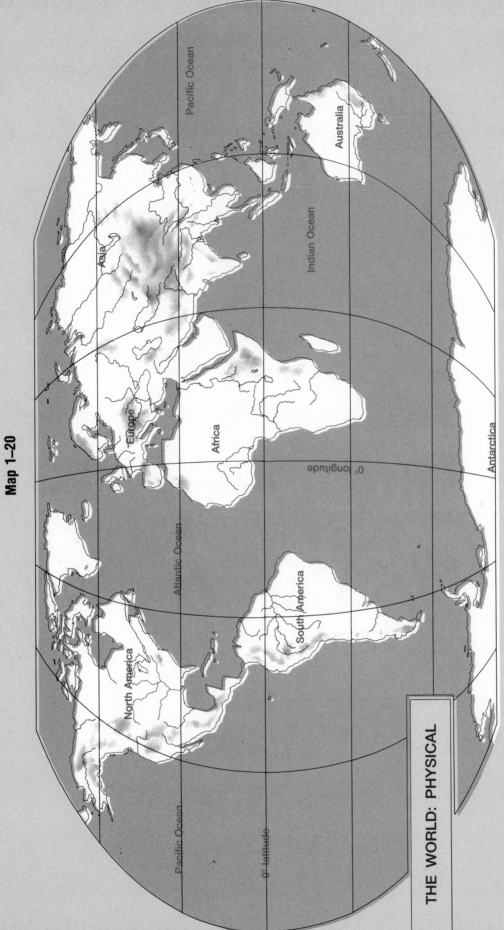

THE WORLD: PHYSICAL

UNIT 2   Place: Physical and Human Characteristics

W hatever the location, there are certain things about a place that make it what it is. Every location has certain physical and human features that make it different from any other. This is what geographers mean when they talk about place. They do not mean *where* a place is (location). Instead, they mean what a place is *like*.

Places on the earth have physical features, such as mountains, deserts, lakes, rivers, plants, and animals.

Places on the earth also have human features. Do many people live there, or only a few? What kinds of buildings have the people put up? What do they do for a living? What do they do for fun? What language do they speak? What kind of government do they have?

Both physical and human features of a place can be described in a number of ways. One way would be to write a detailed description of every feature. However, this would take a very long time to read. Geographers often present information about places in ways that are shorter and easier to understand. Maps, graphs, and tables can be used to present information that would be difficult to understand in other forms.

In this unit, you will learn how to read and construct a number of different kinds of maps, graphs, and tables. The information in maps, graphs, and tables is used to describe place—the physical and human features that make a location different from any other.

What an Address

If you get tired of writing down your address, think about the people that live in Krungthep Mahanakhon, which is the capital of Thailand. The full name of their town is actually Krungthep Mahanakhon Bovorn Ratanakosin Mahintharayutthaya Mahadilok pop Noparatratchathani Burirom Udomratchanivetmahasathan Amornpiman Avatarnsathit Sakkathattiyavisnukarmprasit. With 167 letters, this name is enough to drive mapmakers and sign painters crazy.

Elsewhere there are five towns whose names consist of one letter. These are the French village of Y, the Danish village of A, and the Swedish village of Å, the village of U located in the Caroline Islands of the Pacific Ocean, and the Japanese town of Sosei, which is also called O.

Old Map

The oldest known map was made around 4,000 years ago in Babylonia (an area we now know as Iraq). Experts believe that the map probably shows an estate located in a mountain valley. Made on a clay tablet, this is one map that would be hard to fold up and stuff into the glove compartment of a car.

Place: physical and human characteristics is one of the five basic themes of geography. Once we know about weather, and land, and how people live, work, and play, we know a great deal about a place. If we were set down in the middle of a village in the middle of the Sahara, for example, we might not know where we were, but we would know at once that we were not in the United States. The physical and cultural characteristics of that village would tell us that we were somewhere quite different from the United States.

49

OBJECTIVE

Distinguish between types of landforms and bodies of water

TERMS TO KNOW

landform (LAND·form)—feature of the earth's surface

The earth is a very interesting place. Everywhere you turn there is something you have not seen before. Humans seem to have a need to explore, to go places they have never been before. In fact, the travel industry is one of the largest in the world.

The earth would not be so interesting if all places on it looked the same. Your family probably would not travel hundreds of miles to visit a place exactly like your home. The earth does not look the same everywhere because it has many different **landforms**. Landforms are features of the earth's surface, such as mountains and plains.

Using Your Skills

A Read each description of a landform or body of water. Then write the term in the correct place on Map 2–1.

bay—part of a body of water that extends into the land

canyon—a deep, narrow valley with steep sides

cape—curved or pointed part of a coastline that sticks out into the water

coast—land beside a sea or an ocean

delta—land built up at the mouth of a river

gulf—part of a body of water that extends into the land; larger than a bay

island—body of land completely surrounded by water

isthmus—narrow piece of land connecting two larger pieces of land

lake—body of water completely surrounded by land; usually fresh water

mountain—raised area of land, usually with steep sides; the highest landforms

mouth of a river—place where a river empties into a larger body of water

peninsula—narrow piece of land sticking out into water and surrounded by water on three sides

plain—flat or gently rolling land, usually not far above sea level

plateau—flat or gently rolling land higher than the surrounding land

river—large stream of water that flows through land

tributary—a small stream of water that flows through the land feeding into a river

valley—low land between hills or mountains

Map 2–1 **LAND AND WATER FORMS**

Lesson 2   Reading Elevation Maps·

OBJECTIVE

Use elevation maps to gain information about the physical characteristics of a place

TERMS TO KNOW

altitude (AL·tuh·tood)—elevation

elevation (el·uh·VAY·shuhn)—height

elevation map (el·uh·VAY·shuhn map)—map that shows elevation

relief (ruh·LEEF)—difference in elevation between places

sea level (see LEV·uhl)—average height of water in the world's oceans

D oes it bother you to climb a ladder? Are you afraid to look down from the top of stairs or buildings? If so, you already know something about **elevation,** or height.

When you are standing on the top rung of a ladder, you may be four feet above the ground. However, your elevation will not be four feet, unless you happen to be standing on a ladder at the beach. It is important to remember that all elevation is measured from **sea level.** Sea level is the average level of the water in the world's oceans, or zero feet. You may be four feet above the ground, but your elevation may be 2,000 feet above sea level. Or, in a few spots on the earth's surface, your elevation could be 100 feet or more *below* sea level.

The Importance of Elevation

Knowing your elevation can sometimes be a matter of life and death. Suppose you are flying a plane at an

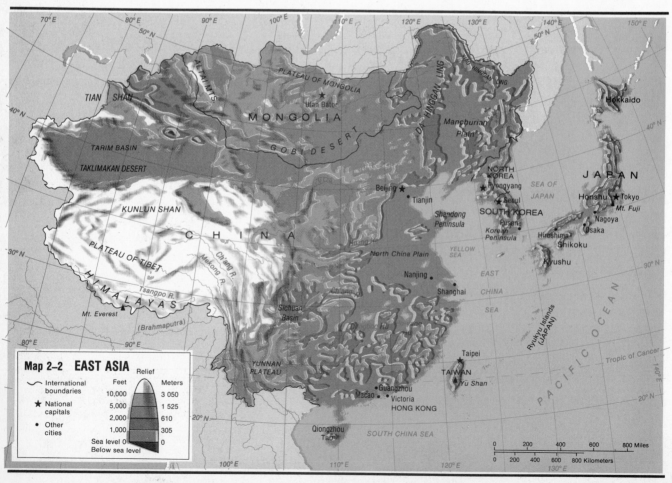

Map 2–2   EAST ASIA

Symbol		Feet	Meters
⌒	International boundaries	10,000	3 050
★	National capitals	5,000	1 525
•	Other cities	2,000	610
		1,000	305
		Sea level 0	0
		Below sea level	

Relief

0   200   400   600   800 Miles

0   200   400   600   800 Kilometers

altitude of 10,000 feet. *Altitude* is another word for elevation. Ahead of you is a mountain. Your map says the elevation of the top of the mountain is 12,500 feet. What will happen unless you do something?

Elevation is an important characteristic of a place. In general, the greater the elevation, the cooler the climate will be. The difference in elevation between places, or **relief,** is also important. If moisture-bearing winds blow from a low area to a higher area, more rain will fall at the higher elevations. Streams flow from higher land to lower land.

Showing Elevation on Maps

Look at Map 2–2 of East Asia on page 52. Shading is used on the map to show the elevation of the land. This kind of map is often called a relief map or **elevation map.**

A map cannot show the elevation of every single spot. Areas are grouped together. For example, on the map of East Asia, all areas with an elevation between sea level and 1,000 feet are shaded the same. Within that area may be hills and valleys, but no hill will be higher than 1,000 feet; and no valley will be lower than sea level.

Elevations on this map are grouped in the following way. All elevations below sea level are shaded the same. Elevations from sea level to 1,000 feet are shaded alike. So are elevations from 1,000 feet to 2,000 feet, 2,000 feet to 5,000 feet, 5,000 feet to 10,000 feet, and 10,000 feet and above.

What part of the map tells you what each kind of shading means? Practice reading the legend for a moment. What does this kind of shading mean � ?

Above what elevation are all areas shaded the same no matter how great the elevation?

Using Your Skills

A Write the meaning of each word.

1. elevation— <u>height above sea level</u>

2. altitude— <u>height above sea level</u>

3. relief— <u>difference in elevation</u>

4. sea level— <u>average level of water in the oceans</u>

5. elevation map— <u>map which shows the elevation of the land</u>

B Use Map 2–2 of East Asia on page 52 to answer these questions.

1. What is the elevation of the South China Sea? <u>zero feet, or sea level</u>

2. Look at the part of the legend on the map that shows relief.

    a. Where is the shading for low elevations located? <u>at the bottom of the legend</u>

    b. Where is the shading for high elevations located? <u>at the top of the legend</u>

3. Find the North China Plain on the map. What is its elevation? <u>sea level to 1,000 feet</u>

4. What is the elevation of the land around the Sichuan Basin to the north, south, and east?

    <u>2,000 feet to 5,000 feet</u>

5. One area on the map has an elevation below sea level. Give the location of that area using latitude and

    longitude. <u>43°N latitude, 89°E longitude</u>

Lesson 3   Using Contours to Determine Elevation

Y ou have read how elevation maps show the height of land over large areas. But suppose you are trying to decide where to put a new road. You need to know the exact elevation of particular places in order to choose the best route. An elevation map would not be very helpful.

Contour Maps

There is a map which does show elevation more exactly. This kind of map is called a **contour map.** A contour map has many contour lines on it. These lines show elevation. Each line on the map joins all the places that have the same elevation.

Look at Example 2–1. It shows how a contour map is made. This drawing is of an island. Look at the part of the drawing marked "Top View." This shows you how the island would look from an airplane. Look at the part of the drawing marked "Side View." This shows you how the island would look from a boat on the water.

Notice the lines on the drawing marked "Side View." These imaginary lines "cut through" the island at different elevations. The elevations are marked on the drawing. The first line is at sea level, or 0 feet. At what elevation is the next higher line? The highest?

Now look at the top view. Imagine yourself in a plane high above the island, looking down. Imagine that you could see where all the lines that "cut through" the island come out. The lines would look like the top view. Each line is called a **contour line,** because it follows the shape, or contour, of the land. *Each contour line joins points with the same elevation.* Each line is numbered to show how high above sea level the points joined by that line are located.

By reading the numbers on contour lines, you can tell how high each line is above sea level. How high above sea level is the highest line in the drawing located?

In the side view of the drawing, the lines are all the same distance apart. Why are the lines not the same distance apart in the top view? The lines are not the same distance apart because they follow the shape of the land. Look at the left-hand side of both the top view and the side view. In the side view, you can see that the island slopes up gently from the sea on the left. In the top view, you can see that the contour lines are far apart on the left side. Now look at the middle of both views. The island rises steeply in the middle, as you can see in the side view. In the top view, the contour lines are close together in the middle of the drawing.

This is one of the most important things you need to remember about contour maps. When the lines on a contour map are close together, the land is steep. When the lines are far apart, the land is flat.

Example 2-1

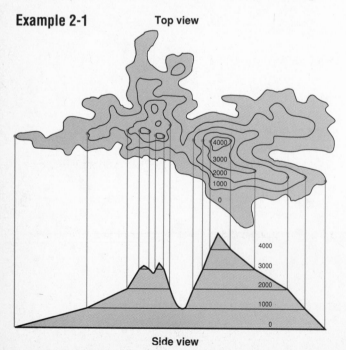

Top view

Side view

54

Reading Contour Maps

When reading a contour map, look for the legend. The legend will tell you whether the contour lines are numbered in feet or meters. It will also tell you how much elevation there is between contour lines. This is called the **contour interval.**

Map 2–3 is part of a contour map of Death Valley, California. Part of Death Valley is below sea level. Can you find the dark contour line that runs across the middle of the map from top to bottom? Near the bottom of the map the line is marked "sea level." That contour line joins the points on the map that are at 0 elevation.

Now look directly to the right of the words *sea level.* Near the edge of the map you will find another contour line that is darker than the others. It is marked "400." That line joins points on the map that are 400 feet *above* sea level. Now look to the left of the words *sea level.* Near the edge of the map on the left you will find a contour line marked "−240." Notice the minus sign in front of the numbers. This means that the line joins points that are *below* sea level.

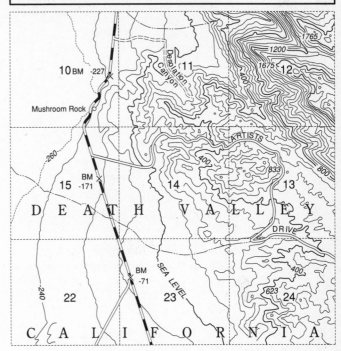

Map 2-3 CONTOUR MAP OF DEATH VALLEY, CALIFORNIA

Contour intervals in feet

Using Your Skills

A **Fill in the blanks with the correct words to complete the following sentences.**

1. Each line on a _____contour_____ map is called a contour line.

2. Each contour line connects points with the same _____elevation_____.

3. When the lines on a contour map are close together, the land is _____steep_____.

4. When the lines on a contour map are far apart, the land is _____flat_____.

5. The amount of elevation between contour lines is called the contour _____interval_____.

6. On a contour map numbered in meters, a contour line marked "150" connects all points on that map that are 150 meters _____above_____ sea level.

7. On a contour map numbered in feet, a contour line marked "-100" connects all points on that map that are 100 feet _____below_____ sea level.

B   Use Map 2–4 of the area around Ithaca, New York, to answer the questions.

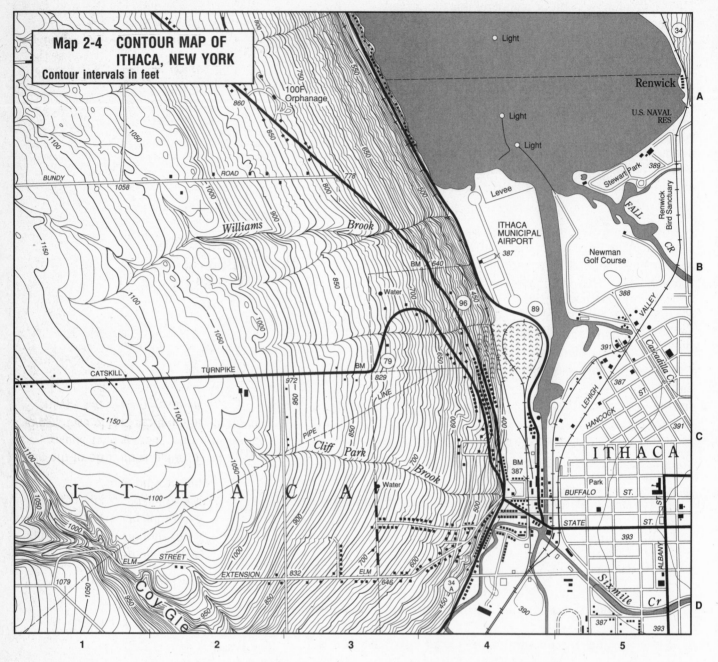

Map 2-4 CONTOUR MAP OF ITHACA, NEW YORK
Contour intervals in feet

1. Is the land around the Newman Golf Course (in cell B-5) flat or steep? How do you know?   It is flat. The

contour lines in that part of the map are far apart.

2. What is the land like just to the west of the Ithaca Municipal Airport? How do you know? (Remember that if

there is no compass rose on a map, north is at the top.)   The land is very steep. The contour lines are very close

together.

3. Find the numbered contour line in the flat area of cell D-4. What does the number on that line mean?   It

means that all the points on that line are 390 feet above sea level.

Lesson 4 Reading Population Density Maps

Think about where you live. Do many other people live nearby? If you were to count all the people who live within half a mile of you in all directions, how many would there be? Ten? Four hundred? Two thousand? Ten thousand?

Human population is not spread evenly over the face of the earth. Some parts of the earth have only a few people, living very far apart. In other places a great many people live very close together.

Population Density Maps

Maps can show where on the earth's surface large

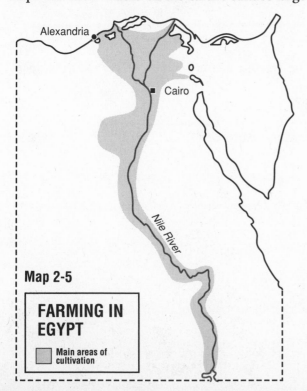

Map 2-5

FARMING IN EGYPT

Main areas of cultivation

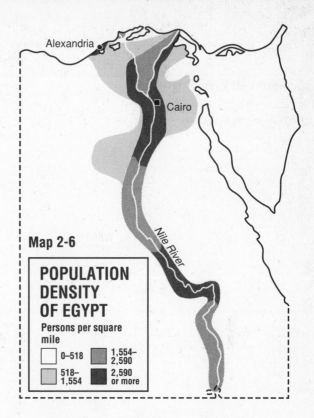

Map 2-6

POPULATION DENSITY OF EGYPT

Persons per square mile

0–518   1,554–2,590

518–1,554   2,590 or more

numbers of people live close together. This kind of map is called a **population density map.** A population density map uses shading (or colors) to show how many people live in each square mile (or square kilometer). Usually, population density maps use lighter shading or color for areas where few people live. Darker color or shading is used for areas where many people live. The legend tells you what each color or shading on the map means.

Comparing Population Density Maps

People tend to live closer together in places where there are more resources to support life. For example, look at Maps 2–5 and 2–6 of Egypt. Map 2–5 shows where farming takes place in Egypt. As you can see, farming takes place along the Nile River.

Now look at Map 2–6. What is the title of the map? The line just under the title, "Persons Per Square Mile," tells you what the numbers beside each symbol in the legend mean. How many people per square mile live in areas with this shading ■ ?

Compare the two maps of Egypt. What conclusion would you draw about the importance of agriculture to the people of Egypt? What conclusion would you draw about the importance of the Nile River?

Using Your Skills

A   **Use Map 2–7 of East Asia to answer the questions.**

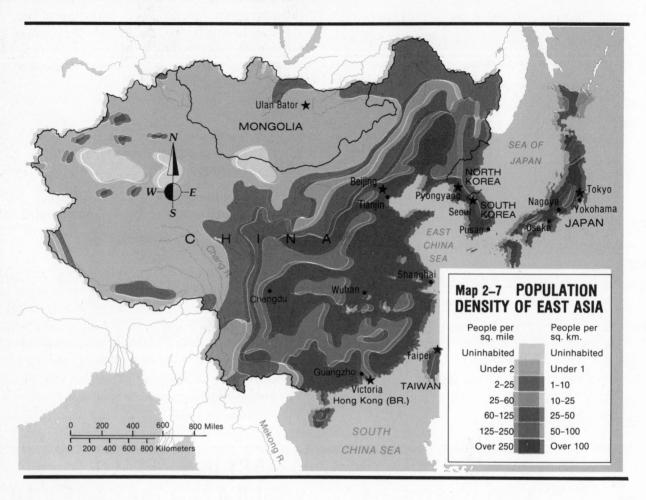

1. What part of East Asia has the highest population density (the greatest number of people per square mile)?

the eastern part

2. What happens to population density in China as one travels from east to west?   The population density goes

down, or gets less.

3. China has more people than any other country. Are there parts of China where no people live? How do you

know?   Yes. There are areas on the map marked "uninhabited."

4. How many people per square mile live in the area around the city of Wuhan?   over 250

5. Look back at Map 2–2 of East Asia. Compare it to the population density map of East Asia. What conclusion

can you draw about the relationship between elevation and population density in East Asia?   Population

density is greater at lower elevations.

OBJECTIVE

Obtain and use information about places from bar graphs

TERMS TO KNOW

bar graph (bar graf)—graph that uses bars to show information

Suppose someone gave you the following paragraph to read and then asked you questions about it:

"Life expectancy varies widely in Latin America. The average person lives 70 years in Mexico, while in Bolivia its 61 and in Haiti the figure is only 46 years. A person in Peru can expect to live 65 years, compared to 77 years in Costa Rica and one year less in Jamaica. Life expectancy in Nicaragua is just over 63 years; in Brazil just over 67 years, and 70 years in Venezuela."

You might have a hard time answering these questions about the paragraph above: In what country do people have the longest life expectancy? The shortest? In which four countries do people have the longest life expectancies? In how many countries can people expect to live for more than 70 years?

Bar Graphs

A **bar graph** presents the same information in another way. A graph is a drawing that shows numbers. A bar graph uses bars to show the numbers.

Bar graphs have a title. The title tells you what the graph is about. Look at Graph 2-1. It is titled "Life Expectancy in Selected Countries in Latin America and the Caribbean."

Two kinds of information are usually shown on bar graphs. One kind of information is shown *across* on the graph. The other kind is shown *up and down*. When you read a bar graph, always look for labels. They will tell you what information is being shown. Be especially careful if a label reads something like "millions of people." This tells you that you must add the word *million* to any number you read.

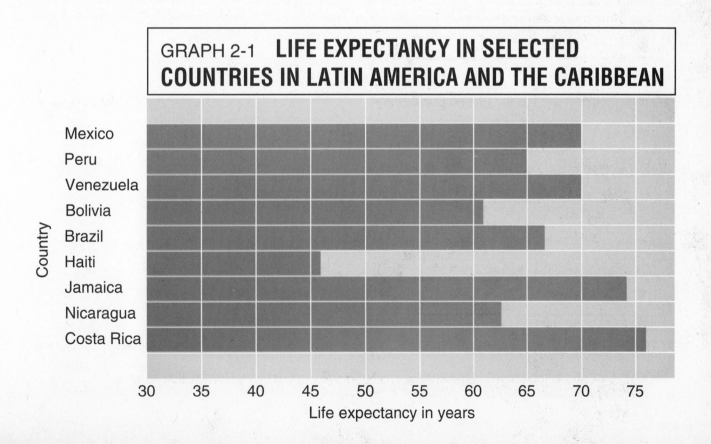

GRAPH 2-1   LIFE EXPECTANCY IN SELECTED COUNTRIES IN LATIN AMERICA AND THE CARIBBEAN

Country: Mexico, Peru, Venezuela, Bolivia, Brazil, Haiti, Jamaica, Nicaragua, Costa Rica

Life expectancy in years: 30, 35, 40, 45, 50, 55, 60, 65, 70, 75

The names of countries are shown *up and down* on Graph 2–1 "Life Expectancy in Selected Countries in Latin America and the Caribbean." The label "Country" beside the names of the countries tells you this. What kind of information is shown *across* on the graph? What tells you this?

Using Bar Graphs

Bar graphs are "read" at the ends of the bars. Look at the bar for Peru on Graph 2-1 "Life Expectancy in Selected Countries in Latin America and the Caribbean." The end of the bar falls on the line marked 65 at the bottom of the graph. This means that the life expectancy in years in Peru is 65. In many cases, the bar will not end on a line. For example, look at the bar for Nicaragua. It ends beyond the line for 60, but not quite at the line for 65. When bars end between lines, you must estimate the number the bar stands for. What number would you estimate the bar for Nicaragua stands for?

Bar graphs are very useful for comparing numbers. For example, it is easy to see in which country in Latin America people have the longest life expectancy because the bar for Costa Rica is longer than all the others. It is also easy to see that the bars for Mexico and Venezuela are the same, and are also longer than most others.

Bar graphs make it easy to answer questions such as this one: How much longer can a person expect to live in Venezuela than in Peru? To answer a question like this, look at the bars for the two countries. Decide what number each bar stands for. Subtract the smaller number from the larger. In this case, you would think, *The bar for Venezuela ends at 70. The bar for Peru ends at 65. Subtract 65 from 70. The answer is 5 years.*

Using Bar Graphs to Compare Information

Bar graphs can present information about two groups of things *at the same time.* This makes them especially useful for comparing information. Look at Graph 2–2 below titled "Life Expectancy at Birth in the United States."

Notice that in this graph, the bars run up and down. (It makes no difference which way the bars go. Often they are arranged in the way that will give the most room for printing the labels for the bars.) Also notice that bars are shown in pairs on this graph. The legend tells you that in each pair, one bar stands for males, and the other stands for females. When you see a bar graph with pairs of bars, or one bar divided into parts, look for a legend.

Look at Graph 2-2 "Life Expectancy at Birth in the United States." You can see that U.S. women can expect to live longer than men. What on the graph tells you this? How much longer could a woman born in 1940 expect to live than a man born the same year? How did you arrive at your answer? How can you tell that life expectancy has been increasing between 1920 and 1994?

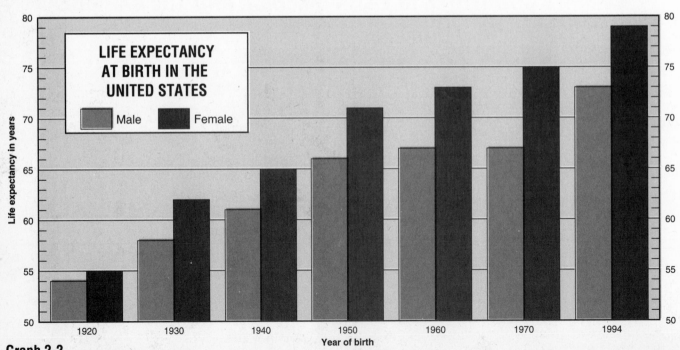

Graph 2-2

Using Your Skills

Use Graph 2–3 to answer the following questions.

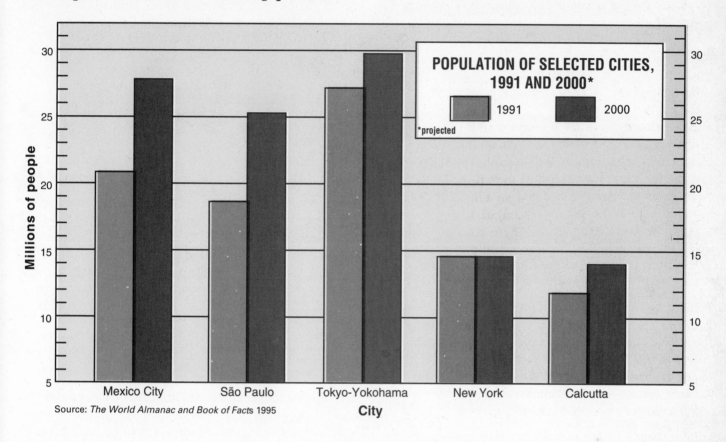

POPULATION OF SELECTED CITIES, 1991 AND 2000*

1991   2000

*projected

Millions of people

Mexico City   São Paulo   Tokyo-Yokohama   New York   Calcutta

City

Source: *The World Almanac and Book of Facts* 1995

1. What is the title of this graph?   Population of Selected Cities, 1991 and 2000

2. Did 15 people or 15 million people live in Mexico City in 1980? How do you know?   Twenty-one million

   people lived in Mexico City in 1991. The label that goes up and down says "millions of people."

3. Which city had the most people in 1980?   Tokyo-Yokohama

4. Which city is expected to have the most people in 2000? How many people is it expected to have?

   Tokyo-Yokohama is expected to have over 29 million in 2000.

5. Which city is expected to be about the same size in 2000 as it was in 1980?   New York

6. About how many more people are expected to live in Calcutta in 2000 than lived there in 1980?

   2 million

7. List the five cities from largest to smallest in 1980.   Tokyo-Yokohama, Mexico City, São Paulo, New York, Calcutta

8. List the five cities from largest to smallest in 2000.   Tokyo-Yokohama, Mexico City, São Paulo, New York, Calcutta

Lesson 6   Reading Line Graphs

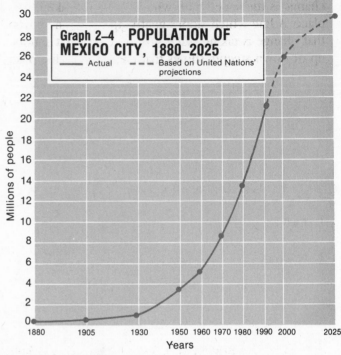

Graph 2–4   **POPULATION OF MEXICO CITY, 1880–2025**

——— Actual   – – – Based on United Nations' projections

Source: *World Almanac 1995*

Often we need to show two kinds of information on a graph: an amount and time. For example, you might wish to show how you grew taller as you became older. This kind of information is best shown on a **line graph.**

A line graph is made up of a grid. Lines going across and up and down cross each other. One kind of information, usually an amount, is shown on the lines going up and down. The other kind of information, usually time, is shown on the lines going across.

Using Line Graphs to Show Trends

Line graphs are especially useful for showing how an amount changes over time. How the amount changes is called the **trend.** If the line on the graph goes up, we say the trend is up. If the line on the graph goes down, we say the trend is down. If the line shows little change up or down, we say the trend is flat.

Use the steps below to read Graph 2–4 titled "Population of Mexico City, 1880–2025."

1. Find out what kind of information is shown on the graph. Look for the title of the graph. It will tell you what the graph is about. Often the title will also tell you what period of time the graph covers. What is the title of this graph? What time period does the graph cover?

Notice the legend below the title. This explains the meaning of the two kinds of lines shown on the graph. If only one line is shown on a graph, no legend is needed. What does the legend tell you the solid line on the graph means? What does the dashed line mean?

2. Study the labels on the graph. The labels are the words that tell you what the numbers on the graph stand for. Look at the numbers on the left side of the graph. The number 0 is at the bottom of the graph. Above it are numbers up to 30. What does the label say these numbers stand for? Now look at the numbers across the bottom of the graph. Next to 0 on the left side is the number 1880. At the far right is the number 2025. What does the label tell you these numbers stand for?

3. Read the line graph. Begin reading the graph from the left. Read across the graph to the right. Notice that there is a dot on the line above each year. To find out how much the dot stands for, read across to the left. For example, the dot for the year 1930 is about halfway between the 0 and the 2 on the left. This tells you that about how many million people lived in Mexico City in 1930? About how many million people lived in Mexico City in 1970? How many million are expected to live there in the year 2000?

4. Draw conclusions from the graph. Observe the line on the graph. Line graphs are so useful because just a glance at one can tell you not only the trend, but also the **rate of change.** The rate of change is the speed with which change is taking place. A line which slopes gently up or down shows that change is taking place slowly. A line that shoots up or down steeply shows that change is taking place rapidly. Is the trend of the graph up or down? Is the population of Mexico City increasing or decreasing? What was the rate of change in the population of Mexico City between 1880 and 1930? What was the rate of change between 1930 and 1980? What is expected to happen to the rate of change between the year 2000 and 2025?

Using Your Skills

A **Explain the meaning of each term below.**

1. trend   Trend is the direction of change shown on a line graph. The trend can be up, down, or flat.

2. rate of change   Rate of change is the speed with which change is taking place. Rate of change is indicated by the slope of the

line. Little slope indicates a slow rate of change. Steep slope indicates a rapid rate of change.

B **Use the information in the paragraph below to complete the line graph (Graph 2–5) which has been started for you.**

The estimated world population in the year 1800 was 1 billion. In 1930 the 2 billion mark was reached. By 1960 there were 3 billion people in the world. As of 1975 there were 4 billion. The number reached 5 billion in 1987 and should reach 6 billion by 1998. The number may reach 7 billion by 2010.

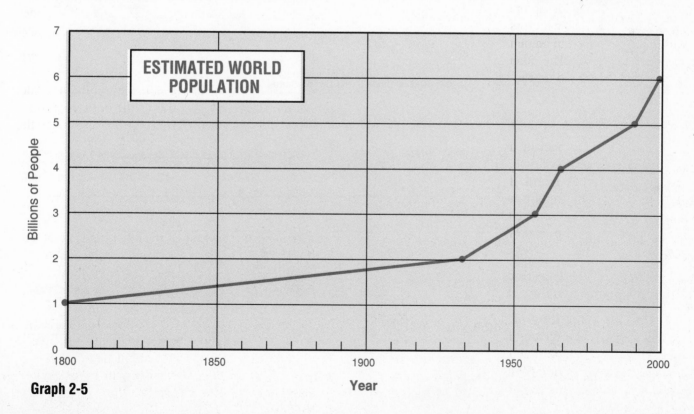

Graph 2-5

C   **Use Graph 2–6 to answer the questions.**

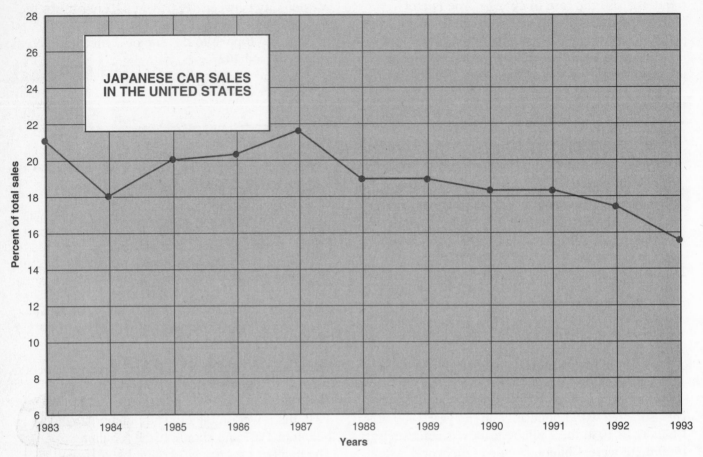

Source: *World Almanac 1995*

Graph 2-6

1. What is the title of the graph?   Japanese Car Sales in the United States

2. For what years are sales given?   1983–1993

3. What do the numbers at left, from 6 to 28, stand for?   the percent of total sales made up by Japanese car sales in the

United States

4. What do the numbers across the bottom of the graph stand for?   years

5. About what percent of total car sales in the United States were Japanese car sales in 1987?   about 21 percent

6. About what percent of total car sales in the United States were Japanese car sales in 1990?   about 18 percent

7. What was the trend in Japanese car sales in the United States between 1987 and 1993?   The trend was down.

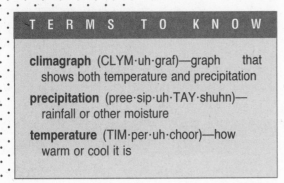

OBJECTIVE

Obtain information about places from climagraphs

TERMS TO KNOW

climagraph (CLYM·uh·graf)—graph that shows both temperature and precipitation

precipitation (pree·sip·uh·TAY·shuhn)—rainfall or other moisture

temperature (TIM·per·uh·choor)—how warm or cool it is

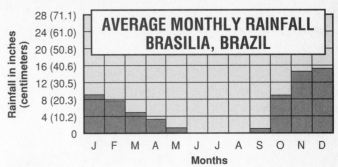

Graph 2-7   *The Weather Almanac, 3rd ed.* Detroit: Gale Research Co., 1981.

When you are making plans to do something outdoors, what are the two things about the weather you are most interested in knowing? You will want to know how warm (or cool) it will be. You will also want to know if it will be wet or dry. **Temperature** (how warm or cool it is) and **precipitation** (rainfall or other moisture) are the two most important things to know about weather and climate.

Using Temperature/Rainfall Graphs

Information about temperature and precipitation can be shown on graphs. Look at Graph 2–7 at the top of this page. The title tells you that this graph shows the average monthly rainfall for Brasília, Brazil.

Read the label to find out what the numbers on the left side of the graph mean. It is very important when

reading climagraphs to notice whether temperatures are given in degrees Fahrenheit (°F) or degrees Celsius (°C). You should also note whether rainfall amounts are given in inches (in.) or centimeters (cm). Some graphs give both.

The letters across the bottom of the graph stand for the months of the year. The first letter on the left is *J.* It stands for January, the first month.

Notice that the names of several months begin with the same letter—*J, M,* or *A.* If you are not sure which month a letter stands for, begin with January at the left and say the names of the months in order, pointing to each in turn. Stop when you get to the name of the month you are interested in reading.

In which month does Brasília receive the most rainfall? In which months does it receive no rainfall?

Information about the average monthly temperature in Irkutsk, Russia, is shown on Graph 2-8 below. Notice that this graph also shows the months of the year beginning with January on the left and ending with December on the right. What is the warmest month in Irkutsk? What is the average temperature in Irkutsk in October?

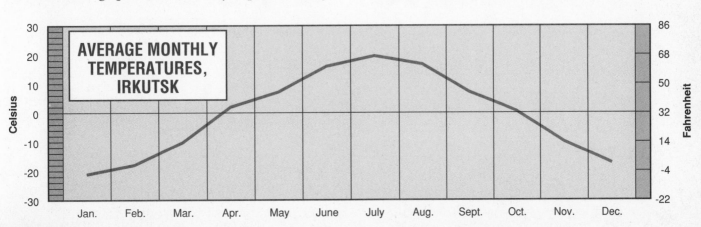

Graph 2-8

65

Using Climagraphs

Information about both temperature and rainfall can be combined on one graph. This kind of graph is called a **climagraph.** A climagraph can look like a bar graph and a line graph put together. Look at Graph 2–9.

The bars across the bottom of the graph show the average monthly rainfall. The numbers on the *right* side of the graph are used to read the amounts on the bars. Do the bars stand for inches of rainfall, or centimeters? What is the average monthly rainfall in Jacksonville in June? In which month does the most rain fall?

Temperature on the climagraph is shown by the line on the graph above the rainfall bars. The numbers on the *left* side of the graph are used to read the temperature line. Are temperatures given in degrees Fahrenheit, or Celsius? What is the average monthly temperature in Jacksonville in December? In which month is the temperature highest?

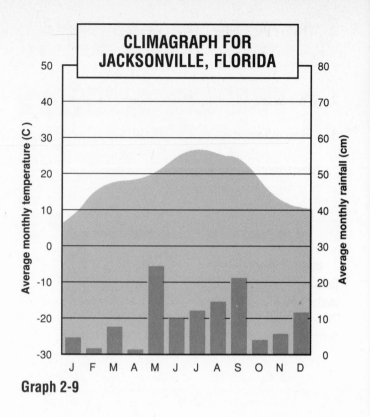

Graph 2-9

Using Your Skills

A   Use Graph 2–10 to answer the questions on page 67.

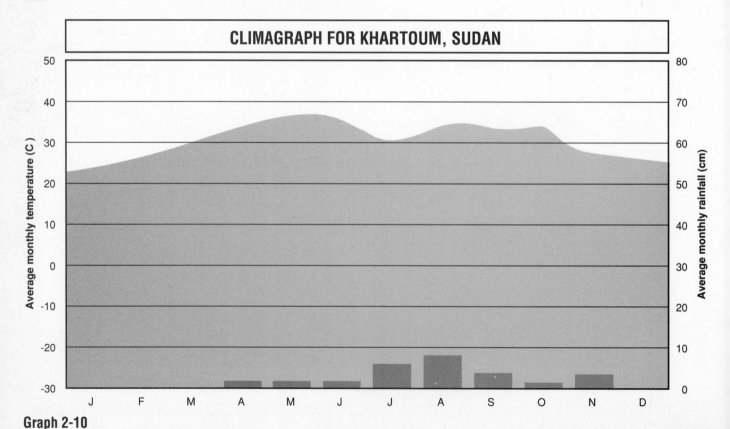

Graph 2-10

1. What is the highest average monthly rainfall for Khartoum? In which month does it occur?   8 centimeters;

August

2. What is the lowest average monthly temperature in Khartoum? The highest?   lowest—22°C; highest—35°C.

3. Using information about rainfall and temperature from the climagraph, how would you describe the climate of

Khartoum in words?   Suggested answer: The climate is hot and dry throughout the year. Little or no rain falls from December

through June. July, August, and September are the wettest months. The hottest months are from March through October.

B   **Construct a climagraph for Cherrapunji, India, using the information and the blank graph.**

Rainfall in Cherrapunji amounts to about 1 inch in January, 2 inches in February, 7 inches in March, 24 inches in April, 67 inches in May, 115 inches in June, 96 inches in July, 71 inches in August, 46 inches in September, 17 inches in October, 2 inches in November, and 1 inch in December. The average temperature in January is 53°F. For the rest of the months of the year in order, the figures are: 58°, 62°, 66°, 67°, 68°, 69°, 69°, 69°, 67°, 61°, 55°.

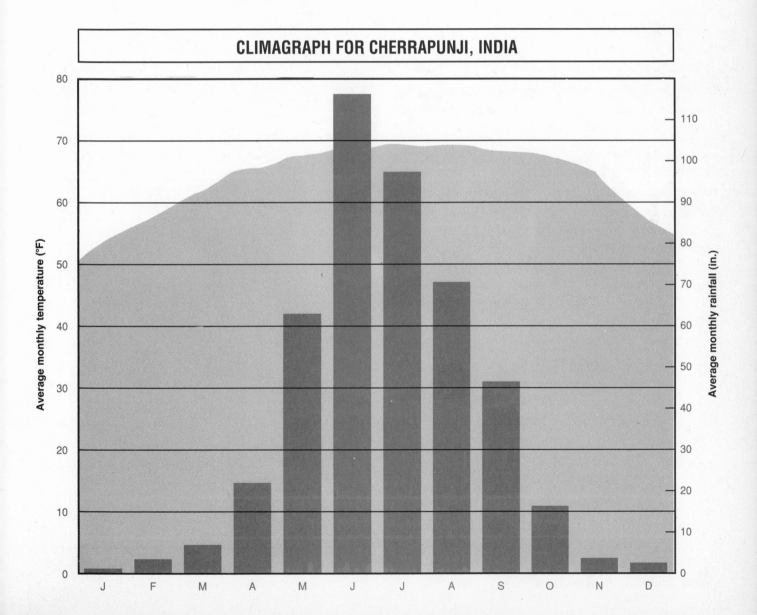

CLIMAGRAPH FOR CHERRAPUNJI, INDIA

Lesson 8   Reading Circle Graphs

When you have pie for dessert, do you try to get the biggest piece? You can tell just by looking if one piece is larger than the others. One quick glance and you know which piece you want!

The kind of graph called a **pie graph** is shaped like a pie. Pie graphs are also called **circle graphs.** As you might guess, circle graphs are round.

Using Circle Graphs

Circle graphs are used for a very special purpose. They are used only when we wish to compare the parts of a whole. For example, suppose the "whole" we wish to consider is the population of Japan. The "parts" we wish to consider are men and women. What part of the population of Japan is made up of men? What part is made up of women? Look at Graph 2–11.

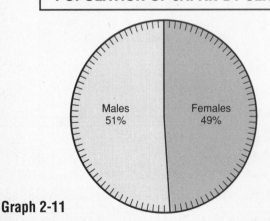

POPULATION OF JAPAN BY SEX

Males 51%   Females 49%

Graph 2-11

The circle graph has been divided into two parts. One shows what part of the total population is made up of men. The other shows what part is made up of women. Which part is bigger, the part for men, or the part for women?

You know that there are 100 cents in a dollar. A dollar is divided into 100 parts. Circle graphs are also divided into 100 parts. Each part is called 1/100, or 1 percent. The symbol for percent is %. The expressions 1/100, 1 percent, and 1% all mean one part out of one hundred.

Reading Circle Graphs

Any time we wish to compare the parts of a whole on a circle graph, first we must divide the whole into 100 parts. Then we figure out how many of those 100 parts belong in each category of things being compared. For example, suppose we wish to compare the population of men and women in India on a circle graph. If 52 percent of the people in India are men, 52 of the 100 parts of the circle graph must be shaded to stand for men. How many parts are left out of the 100? These 48 parts must be shaded in to stand for women. Graph 2–11a shows how this would look.

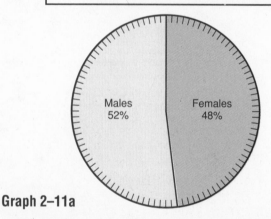

POPULATION OF INDIA BY SEX

Males 52%   Females 48%

Graph 2–11a

When you are reading a circle graph, follow the same steps you would in reading any other graph. First, read the title. Second, look for any other information which will help you read the graph. There may be a date to show what year the information was gathered. A legend may be used to show what the different divisions of the graph stand for. Third, read any labels on the parts of the graph. Often the parts will be labeled with a name and a number to show what percent of the whole the part is.

Using Your Skills

A Use Graph 2–12 to answer the questions.

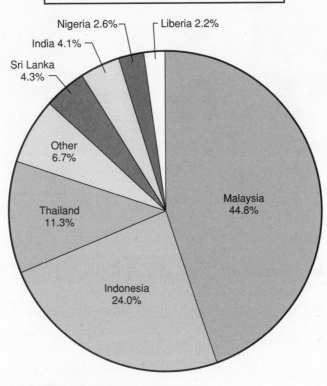

WORLD RUBBER PRODUCTION

Nigeria 2.6%
Liberia 2.2%
India 4.1%
Sri Lanka 4.3%
Other 6.7%
Malaysia 44.8%
Thailand 11.3%
Indonesia 24.0%

Graph 2-12

1. What is the title of this graph?   World Rubber Production

2. Which country produces the most rubber?   Malaysia

3. Which country produces about the same amount of rubber as Nigeria?   Liberia

4. What other two countries produce about the same amount of rubber each?   Sri Lanka and India

5. What percentage of the world's rubber do the countries of Malaysia and Liberia combined produce?

   47%

6. Do the countries of Malaysia, India, and Nigeria combined produce more rubber than the rest of the world combined?   Yes   What percentage of the world's rubber do they produce?

   51.5%

7. What does the label "Other" mean on this graph?   It represents all other countries that also produce rubber.

B Use Graphs 2–13 and 2–14 to compare religious membership in the United States and Canada.

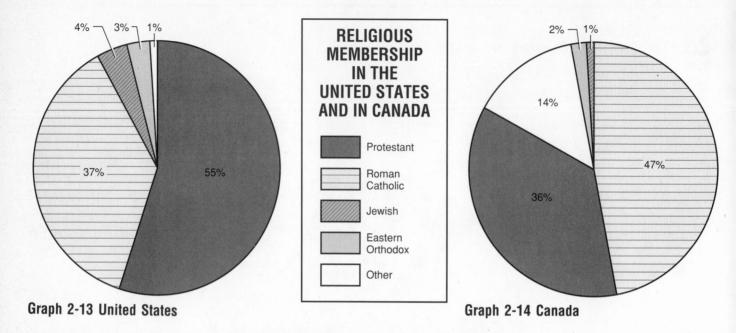

Graph 2-13 United States

RELIGIOUS MEMBERSHIP IN THE UNITED STATES AND IN CANADA

- Protestant
- Roman Catholic
- Jewish
- Eastern Orthodox
- Other

Graph 2-14 Canada

1. In which country are Protestants the largest group?   the United States

2. In which country are Roman Catholics the second-largest group?   the United States

3. In which country are Protestants the second-largest group?   Canada

4. Which country has the smaller percentage of Jewish people?   Canada

C Use the information and blank circle graph to make your own circle graph.

The population of the United Arab Emirates in 1980 could be divided into three groups. Persons 0 to 14 years of age were 29 percent of the population. Persons 15 to 64 were 69 percent. Persons 65 and over were 2 percent.

POPULATION OF THE UNITED ARAB EMIRATES BY AGE, 1980

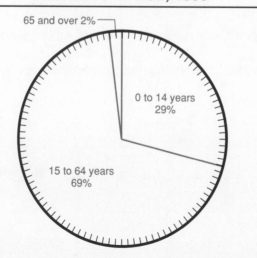

65 and over 2%

0 to 14 years
29%

15 to 64 years
69%

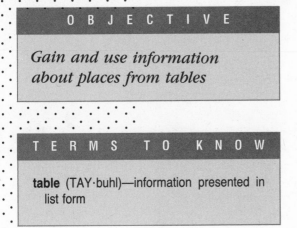

O B J E C T I V E

Gain and use information about places from tables

T E R M S   T O   K N O W

table (TAY·buhl)—information presented in list form

Getting Information From Tables

Usually, of course, tables have more than just two columns of information. Look at Table 2–1. To read a table, follow these steps. First, read the title of the table. What is the title of this table? Second, look for a legend, or for any labels which will help you understand information in the table. For example, the label in parentheses just below the title of this table tells you that the figures in the table are given in millions of U.S. dollars. This means that you must add six zeros to each number when you read it. The labels down the lefthand side of the table are the names of countries. The labels across the top of the table are years.

Third, find information by using the table like a map grid. For example, to find how many dollars were spent by tourists in Singapore in 1991, first find Singapore in the lefthand column. (Names in this column will usually be in alphabetical order.) Place your left index finger on it. Then find 1991 in the row at the top of the table. Place your right index finger on it. Move your left hand straight across, and your right hand straight down, until your index fingers meet. The information you are looking for is in this cell of the table. You add six zeros to the number and read: $4,497,000,000 (four billion four hundred ninety-seven million dollars).

Y ou have probably looked up friends' telephone numbers in the telephone book. The telephone book is a kind of **table.** A table is a list of information presented in a form which can be read easily.

Tables are set up in columns and rows. Columns go up and down the page. Rows go across. In a telephone book, there are two columns for each listing. The person's name is in one column. The telephone number is in the second column. One name and one number make up one row. To find a number, first find the person's name. Then move across the row to the number.

TOURISM IN SOUTHEAST ASIA (in millions of U.S. dollars)				
	1989	1990	1991	1992
Indonesia	1,628	2,153	2,515	2,729
Malaysia	1,038	1,667	1,530	1,768
Philippines	1,465	1,306	1,281	1,674
Singapore	3,307	4,719	4,497	5,204
Thailand	3,753	4,326	3,923	4,829

Source: *UN Statistical Yearbook* © 1994

Table 2-1

Using Your Skills

 A   Use Table 2–2 to answer the questions.

SOME INFORMATION ABOUT THE CONTINENTS OF THE WORLD

Name	Population*	Land Size (sq. mi.)	Land Size (sq. km)
Africa	700,000,000	11,688,000	30,271,920
Antarctica	—	5,100,000	13,209,000
Asia	3,392,000,000	16,999,000	44,027,410
Australia	18,000,000	2,966,000	7,681,940
Europe	728,000,000	4,017,000	10,404,030
North America	449,000,000	9,366,000	24,257,940
South America	311,000,000	6,881,000	17,821,790

* Population is estimated to the nearest million. Antarctica has no permanent population.   Source: *1994 World Population Data Sheet*

Table 2-2

1. What is this table about?   <u>the continents of the world</u>

2. What information is listed in the lefthand column of the table?   <u>the names of continents</u>

3. What three pieces of information are given for each continent?   <u>population, land size in square miles, and land size</u>

   <u>in square kilometers</u>

4. Why is no population figure given for Antarctica? How do you know?   <u>Antarctica has no permanent population. This</u>

   <u>information is given in a note at the bottom of the table.</u>

5. What is the land size of Europe in square miles?   <u>4,017,000</u>

6. What is the population of Asia?   <u>3,392,000,000</u>

7. Which continent has the largest land size?   <u>Asia</u>

8. Which continent has the smallest population (other than Antarctica)?   <u>Australia</u>

B Choose five people you know. Gather information about them and make a table to display the information under these headings: name, age, favorite food, favorite game, and favorite musical group.

PEOPLE I KNOW

Name	Age	Favorite Food	Favorite Game	Favorite Musical Group
		Answers will vary.		

C In the space below, use the information in the table in exercise B to plan a party. What foods will you serve? What games will you play? What music will you have?

Answers will vary.

Lesson 10 Reading Population Pyramids

OBJECTIVE

Gain information about places from population pyramids

TERMS TO KNOW

population bulge (pahp·yuh·LAY·shuhn buhlj)—people who make up a large group in the population

population pyramid (pahp·yuh·LAY·shuhn PIR·uh·mid)—graph that shows how the population is divided by sex and age

Have you ever been to a place where there was no one else around the same age as you? Were most of the people younger than you, or older?

The ages of the people who live in a country are very important. For example, if a country has a population made up mostly of very young and very old people, there will be few people of working age. This can cause food shortages as well as other problems. If a country has a great many young children, there may be a need for more schools and teachers when those children reach school age.

Population Pyramids

People who study population have developed a special kind of bar graph to show how the population of an area is divided by age and by sex. This kind of graph is called a **population pyramid.**

The name of this kind of graph comes from the shape called a pyramid. A pyramid is wide at the bottom and narrow at the top. Look at Graph 2–15, the population pyramid for China.

The scale on the side of the graph tells you what age group each bar represents. For example, the bottom bar on the graph stands for people between the ages of 0 and 4 years. The pyramid is made up of bars that go across. Each bar is divided into two parts by a line down the center of the graph. The left side of the bar stands for males in the population. The right side of the bar stands for females. The legend tells you which shading on the bars stands for males and which stands for females.

At the bottom of the pyramid is a scale marked in percent (%). Notice that the scale begins in the middle of the graph. To read the percent of males, you must read from the center outward *to the left*. To read the percent of females, you must read from the center outward *to the right*. To find the total percent of the population in a particular age group, you must *add* the figures for males and females.

Getting Information From Population Pyramids

Practice reading the population pyramid for China. For what year are figures shown? In what age group are the largest number of males? Females? Add the figures for males and females in the 10 to 14 age group. What percent of the population is between the ages of 10 and 14 years? About what percent of the females in China are between the ages of 40 and 44?

Look at the bars for people in China between the ages of 15 through 19 and 20 through 24. These bars are longer than any of the others. This means that there are more people in China between the ages of 15 and 24 than in any other age group. Between 1990 and 1995, this age group got five years older—they were in the 20 through 29 age group. By 2005, these people will be in the 30 through 39 age group. Do you see what will happen to the shape of China's population pyramid as these people grow older? The wide bars will steadily move toward the top of the pyramid.

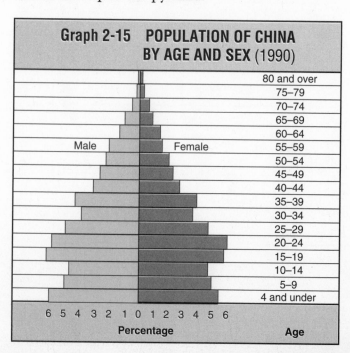

Graph 2-15 POPULATION OF CHINA BY AGE AND SEX (1990)

Male	Female	80 and over
		75–79
		70–74
		65–69
		60–64
		55–59
		50–54
		45–49
		40–44
		35–39
		30–34
		25–29
		20–24
		15–19
		10–14
		5–9
		4 and under

6 5 4 3 2 1 0 1 2 3 4 5 6

Percentage   **Age**

Understanding Population Bulges

The people who make up such a large group in the population are called a **population bulge.** The name comes from the wide bars moving up the pyramid. The United States has such a population bulge. It was created by a high birth rate following World War II. This "baby boom" meant that a much higher number of babies than usual were born between 1945 and 1950. As these people become older, there will be a need for more doctors, hospitals, and services for the elderly.

It is possible to make some guesses about future population growth in a country based on population pyramids. For example, a country with a large population bulge at the bottom of its pyramid now will have a large number of people at the age to have children in a few years. That country may have faster population growth in the future. A country with very little bulge anywhere on its population pyramid may have slow population growth. A country with a bulge at the middle or top of its population pyramid may actually lose population in the future.

Using Your Skills

A   Use the population pyramid of Japan (Graph 2–16) to answer the questions.

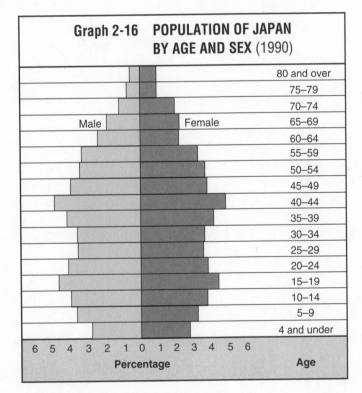

Graph 2-16 POPULATION OF JAPAN BY AGE AND SEX (1990)

1. What age group in Japan has the largest number of people?   40–44

2. What age group in Japan has the second-largest number of people?   15–19

3. The population pyramid of Japan shows two bulges. What age groups make up those bulges?   People ages

10–14, and 15 to 19, 20–24 make up one bulge. People ages 35 to 39, 40 to 44, and 45–49 make up the other.

4. Does the population pyramid show that men or women tend to live longer in Japan? How do you know?

Women tend to live longer. There are more women than men in the older age groups.

B   Use the population pyramids to answer the questions.

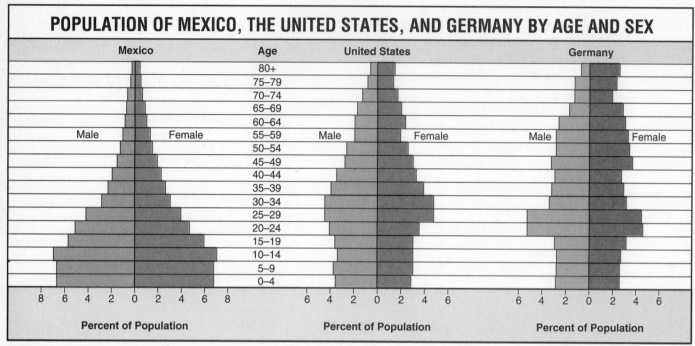

POPULATION OF MEXICO, THE UNITED STATES, AND GERMANY BY AGE AND SEX

Graph 2-17, 2-18, & 2-19

Source: *Demographic Yearbook*, 1991

1. Which country has the highest percentage of people between the ages of 0 and 9?

Mexico

2. In what age groups does the United States have a large population bulge?   ages 20 to 24, 25 to 29, 30 to 34, and

35 to 39

3. Which country has the smallest percentage of people between the ages of 0 and 9?

Germany

4. Which country has the greatest percentage of people at ages 40 and above?   Germany

5. What could happen in Mexico when the people now in the age groups 0 to 4, 5 to 9, and 10 to 14 reach the

age to start having children of their own? Why?   Since there will be a great many people able to have children, a new

population bulge may be created.

6. Will the population of Germany probably grow faster or slower when the people now in the age groups

0 to 4 and 5 to 9 reach the age to have children? Why?   The population will probably grow slower, because there will

be fewer people able to have children.

7. Based on its population pyramid, do you think the population of the United States will grow quickly or slowly?

Why?   The population will probably grow slowly. There are fewer people in the younger age groups to grow up and have children.

Lesson 11 Interpreting a Life Expectancy Map

Reading Life Expectancy Tables

Information about life expectancy can be presented in several different forms. Look at Table 2–3. It shows life expectancy in selected countries in Sub-Saharan Africa.

Table 2–3

Average Life Expectancy

Angola	46
Chad	48
Gabon	54
Gambia	45
Kenya	59
Namibia	59
Nigeria	54
Zimbabwe	56

What country has the highest life expectancy? The lowest? How do these figures compare to life expectancy in the United States?

Information on life expectancy can also be shown on a bar graph. Graph 2–20 is a bar graph with the same information as the table above. Do you see how the graph makes it easier to find the country with the longest or shortest life expectancy?

Using Life Expectancy Maps

Information about the life expectancy of people in an entire region is easily shown on a **life expectancy map.** A life expectancy map uses shading to indicate life expectancy. A legend tells you what the different shades on the map mean.

You probably don't give much thought to how long you can expect to live. Life expectancy is the term we use to describe how long the average person will live. For example, the average person born in the United States in 1994 could expect to live for 76 years. That means that a person born in 1994 could expect to live until the year 2070.

Where in the world you live can have a great deal to do with life expectancy. For example, a person born in a country where there is plenty to eat and good medical attention can expect to live longer than a person born in the same year in a country where food and doctors are in short supply.

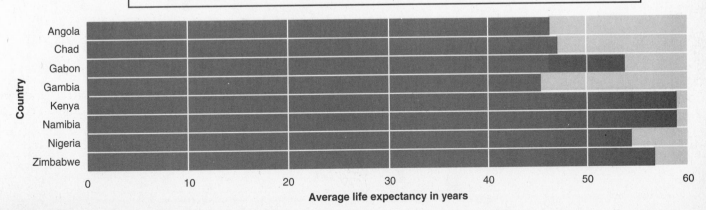

AVERAGE LIFE EXPECTANCY IN SELECTED SUB-SAHARAN COUNTRIES

Country: Angola, Chad, Gabon, Gambia, Kenya, Namibia, Nigeria, Zimbabwe

Average life expectancy in years: 0, 10, 20, 30, 40, 50, 60

A map cannot show life expectancy as exactly as a table or graph can. Instead, age groups are used. For example, one age group might include ages 30 to 35. Another might include ages from 35 to 39, and so on.

You read a life expectancy map in exactly the same way you read any other map which uses shading. First, you read the title of the map. Second, you read the legend to see what each shade used on the map represents. Then you read the information for particular countries.

Using Your Skills

A Use the life expectancy map of Sub-Saharan Africa to answer the questions.

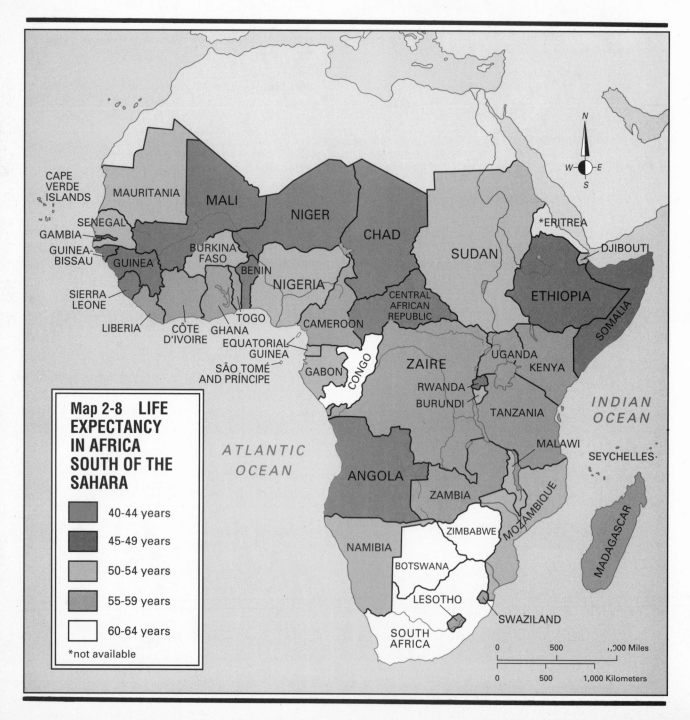

Map 2-8 **LIFE EXPECTANCY IN AFRICA SOUTH OF THE SAHARA**

40-44 years
45-49 years
50-54 years
55-59 years
60-64 years
*not available

1. What is the title of this map?   Life expectancy in Africa south of the Sahara

2. Define life expectancy.   Life expectancy is how long the average person can expect to live.

3. What does this shading ▮ on the map mean?   People in that country can expect to live between 45 and 49 years.

4. How long can people in Mauritania expect to live?   50 to 54 years

5. How long can people in Zaire expect to live?   55 to 59 years

B   Use the map of Sub-Saharan Africa in exercise A to complete Table 2–4.

Life Expectancy in Selected Countries in Sub-Saharan Africa

Country	Life Expectancy (in years)
Angola	45–49
Burkina Faso	50–54
Central African Republic	45–49
Chad	45–49
Congo	60–64
Ethiopia	45–49
Ghana	55–59
Côte d'Ivoire	55–59
Madagascar	55–59
Sudan	50–54
Zambia	55–59

OBJECTIVE

Gain and use information about places from resource maps

TERMS TO KNOW

manufacturing (man·yoo·FAK·cher·ing)— the making of things

natural resource (NACH·er·uhl REE·sohrs) —something that is found on earth

resource (REE·sohrs)—something people use

resource map (REE·sohrs map)—map that shows the things found or produced in an area

W e get information about places in many different ways. One way to learn about a place is by a study of its **resources.** Resources are things people use. Resources can be crops, minerals, animals, or plants. A **resource map** shows the things found or produced in an area. Resource maps do not show exact locations, nor do they show every place where a resource is found. They show the most important places where the resources are found.

Importance of Natural Resources

Natural resources are things that are found on earth. The soil, water, plants, fish, and animals are all natural resources. So are minerals such as gold, iron, and coal. People use natural resources in many ways. One of the most important ways people use the soil is in farming. Many natural resources are used in **manufacturing,** the making of things. For example, iron ore and coal are used to make steel. Steel is then used to make many other things, from cars to school desks.

The resources that are found in a place have a great deal to do with the way people live there. An area with much iron ore and coal may be a center of steel-making. An area with rich soil and plenty of water may be a farming area.

Reading Resource Maps

Reading resource maps uses the skills you have learned in reading symbols and legends. You will use one of two methods to read resource maps, depending on how questions about the map are asked. One kind of question will say, "What natural resources are found in Florida?" To answer this kind of question, you must first find Florida on the map. Then you must see what symbols appear in Florida. Then you must use the legend to see what those symbols mean.

The other kind of question will say, "In which states is gold produced?" To answer this kind of question, you must first use the legend to find what symbol is used for gold. Then you must search the map for all the places where that symbol appears.

Using Your Skills

A **Use Map 2–9 on page 81 to answer the questions.**

1. Use the compass rose to describe in which part of the United States bauxite is found.   the southeastern part of

the United States

2. What natural resource does Canada have that the United States does not?   asbestos

3. What natural resource does the United States have that Canada does not?   bauxite

4. What natural resources are found in and near Hawaii?   fish and forests

5. What natural resources are found in and near Alaska?   oil, natural gas, gold, coal, forests, fish, and shellfish

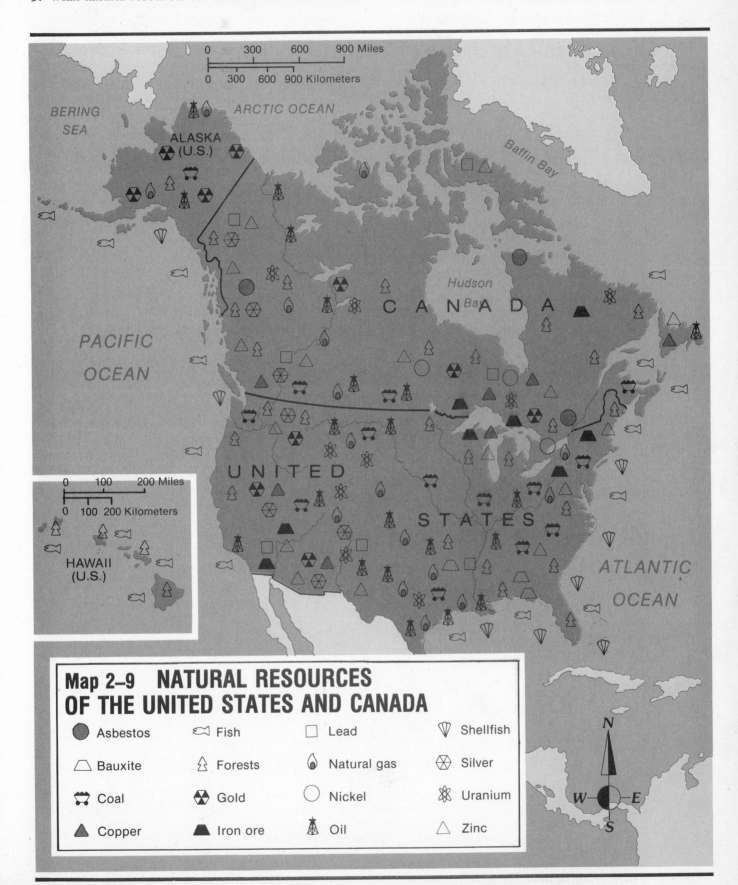

Map 2–9 NATURAL RESOURCES OF THE UNITED STATES AND CANADA

- ⬤ Asbestos
- ⬠ Fish
- ▢ Lead
- ▽ Shellfish
- △ Bauxite
- Forests
- Natural gas
- ⬡ Silver
- Coal
- ⬡ Gold
- ◯ Nickel
- Uranium
- ▲ Copper
- ▲ Iron ore
- Oil
- △ Zinc

B **Use Map 2–10 to answer the questions.**

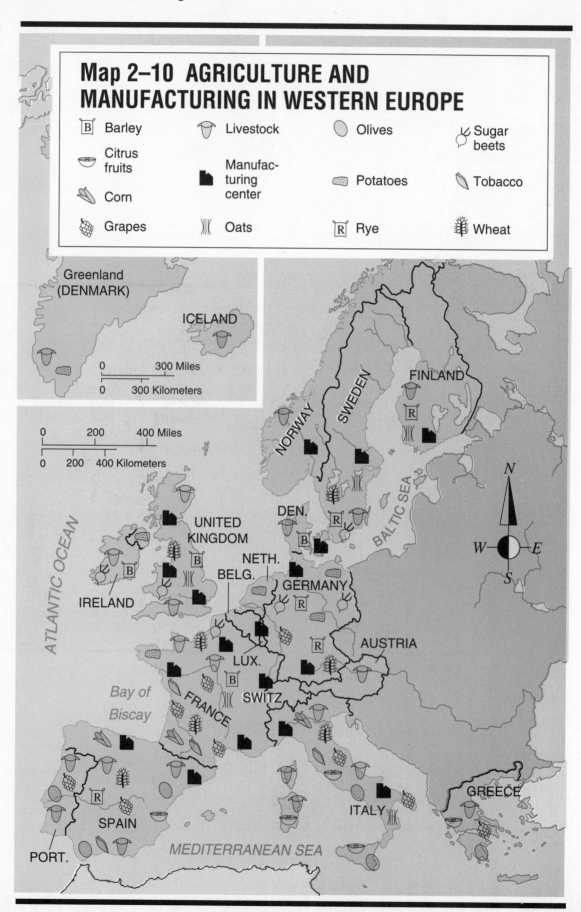

Map 2–10 AGRICULTURE AND MANUFACTURING IN WESTERN EUROPE

1. What resources are found in Greece?   livestock, citrus fruits, grapes, and olives

2. What resources are found in Finland?   livestock, rye, oats, and manufacturing

3. In which countries are protatoes grown?   Greenland, United Kingdom, Netherlands, Germany, and France

4. Which country is shown as having only one resource?   Iceland

5. In which countries are sugar beets grown?   Sweden, United Kingdom, Germany, and France

6. In which countries is corn grown?   Italy, France, and Spain

C   **Use Map 2–11 to answer the questions.**

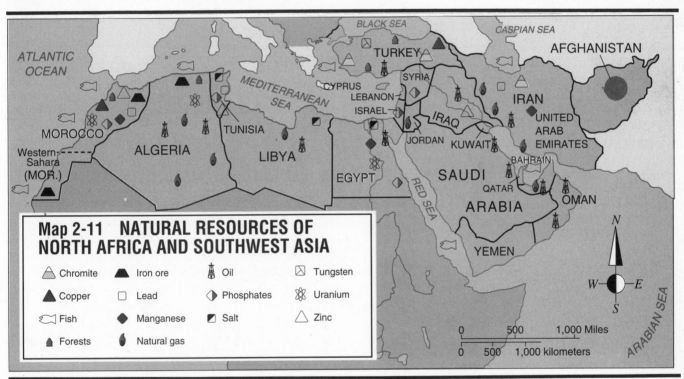

Map 2-11 **NATURAL RESOURCES OF NORTH AFRICA AND SOUTHWEST ASIA**

△ Chromite   ▲ Iron ore   ⛏ Oil   ⊠ Tungsten
▲ Copper   ☐ Lead   ◇ Phosphates   ✳ Uranium
🐟 Fish   ◆ Manganese   ◪ Salt   △ Zinc
🌲 Forests   💧 Natural gas

1. What natural resource is found in Saudi Arabia?   oil

2. What natural resource is found in Afghanistan?   none

3. Which natural resource is found only in Turkey of the countries shown on the map?

tungsten

4. In which countries is zinc found?   Morocco, Tunisia, and Iraq

5. What natural resources are found in Libya?   oil, natural gas, and salt

OBJECTIVE

Gain and use information about places from language maps

Would you like to take a trip around the world? Imagine flying your own plane as you land in Cameroon, Swaziland, Mexico, Peru, Hong Kong, India, Greece, and a dozen other countries. How will you talk to the people in the control towers to get landing instructions? Will you need to brush up on your Swahili, Spanish, Chinese, and Greek before your trip? You won't if you land at airports serving passenger airlines. All landing and takeoff instructions are given in one language around the world: English.

Language is a very important part of human life. We use language to talk with each other in our personal and business lives. Language can bring people together. People who speak the same language may share ideas as well as religious and political beliefs.

Geographers sometimes divide the world into regions based on language. These regions can be shown on maps. Look at Map 2-12 on page 85.

Reading Language Maps

Reading a language map such as this is like reading any other map which uses shading. You must use the legend to determine the meaning of each kind of shading.

In which countries is English the main language? In which countries is English the official language? Which countries teach English in all their elementary schools?

The Spread of the English Language

A language map can lead us to ask some important questions about our world. For example, looking at the map "English as a World Language," one might ask, "Why is English the main language of Canada, the United States, the United Kingdom, Ireland, Australia, New Zealand, and a number of other islands and countries?" You probably already know the answer. These countries were once owned by what we now call the United Kingdom. Part of the United Kingdom is England, and this is where the English language began. People from England spread the English language around the world.

You might also ask why English is the official language in so many countries in Africa. The answer is the same: In most cases, those countries were once owned by the United Kingdom. The rulers took their language with them.

Trade has been another important reason for the spread of the English language. For many years the United States, Canada, and the United Kingdom have been among the richest, most powerful countries in the world. Many of the world's goods are made in these English-speaking countries. Many products from around the world are sold to these countries. The English language has become the language of trade. It is easier for people to buy and sell things when they can speak the same language.

Using Your Skills

A **Answer these questions based on the reading above.**

1. What are two reasons English has become a world language? Many countries were once ruled by English-speaking countries. English has also become the language of trade.

2. Why is language such an important part of our lives? We use language to talk to each other. We use language to share ideas and beliefs. Language helps to bring people together.

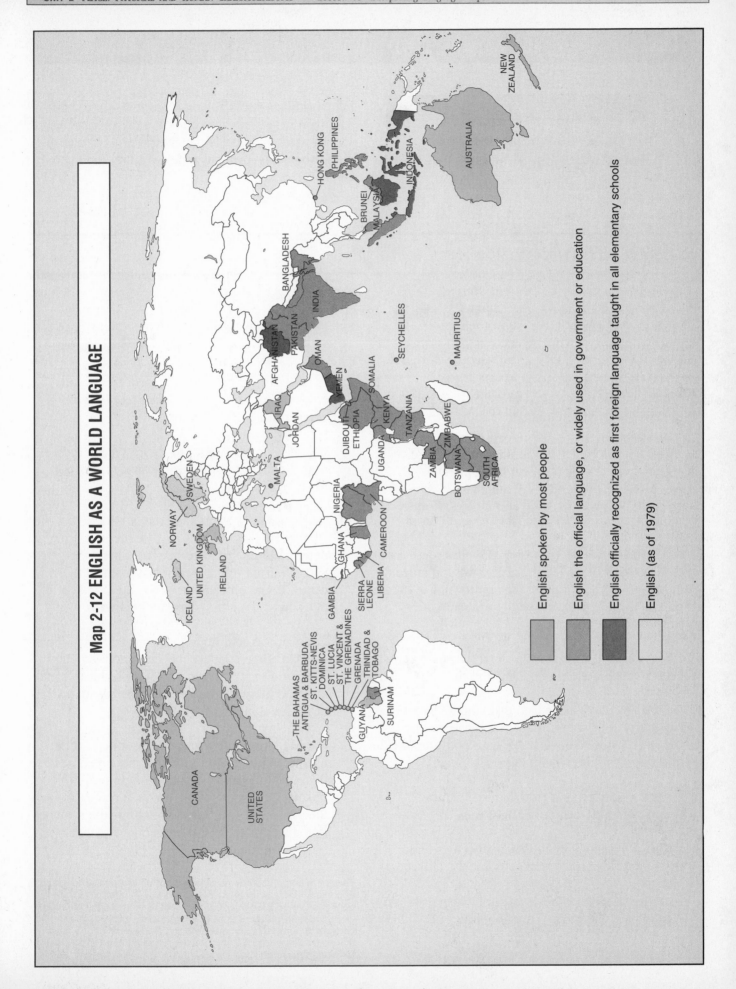

Map 2-12 ENGLISH AS A WORLD LANGUAGE

English spoken by most people

English the official language, or widely used in government or education

English officially recognized as first foreign language taught in all elementary schools

English (as of 1979)

Demonstrate mastery of map and graph skills

This lesson will challenge you to use the skills you have studied in units 1 and 2. If you are not sure how to complete any of the tasks, look back at the lesson or lessons which deal with the skill you need.

Using Your Skills

A **Follow the instructions below to complete Map 2–13.**

1. Read all the instructions before beginning work. What you do in one step may affect a later step, so you need to be aware of all parts of the project before you begin. Draw *lightly* in pencil in case you need to change something later on.

2. Draw in a coastline with Metro City on the west side of a bay. The coastline has been started for you at the top and bottom of the map. Be sure to draw the bay so that Metro City will be on the shore. Name your bay, and label it on the map.

3. Put a small town on the coast 225 miles southeast of Metro City. Put another small town 150 miles northwest of the center of the first town. Use the correct symbol to show the towns. Name the towns and label them.

4. Draw a small lake in cell F-3 and a swamp in cell E-12. Use the correct symbols for each.

5. Draw a river across the land that empties into the bay. Draw at least two tributaries that feed into the river. Draw rivers that lead into both the lake in cell F-3 and the swamp in E-12.

6. Place the symbol for a gold mine 100 miles southwest of the center of Metro City.

7. Locate the capital city in the exact center of the map with the correct symbol. Name the city and label it.

8. The land in cells C-9, C-10, and D-9 is used for farming. Outline this area with a heavy line. Choose a good color to represent farms, and color the area. Place the same color in the correct place in the legend.

9. Use the correct symbol to locate a small town near the center of the farming area. Name the town and label it.

10. Draw highways to connect the farming town, the capital city, and Metro City. Use the map scale to determine the distances between the towns. Measure from the center of one town to the center of the other town. Write the distances in miles between cities along the highways.

11. Connect the gold mine to the nearest city with a railroad. Then connect this city to Metro City with a railroad. Use the map scale to measure the distances between these places, and write the distances along the railroad.

12. Choose a resource that might be found in the northern part of your country, and a resource that might be found in the southern part of your country. Design a symbol for each resource. Draw the symbols on the map. Then draw the symbols in the blank boxes in the legend. Label the symbols in the legend.

Teacher's note: All student maps will not be identical. The shapes and exact locations of many of the required landforms and resource symbols will vary from student to student and must be looked at individually. Before grading this assignment, you should completely familiarize yourself with the student directions found on this page. The annotated map on the following page is provided for example use only.

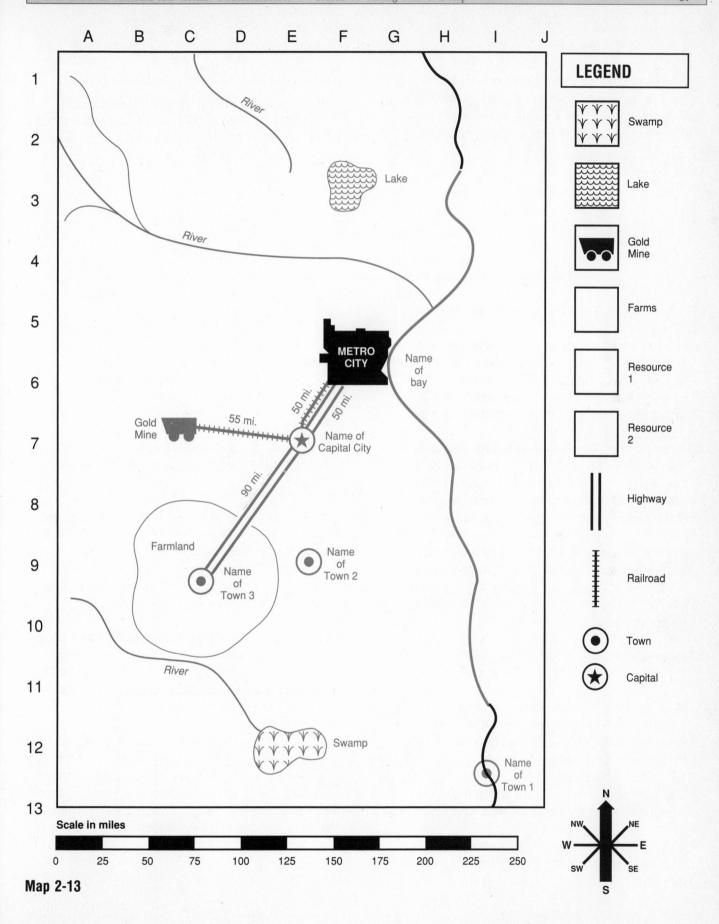

Map 2-13

B   **Using the information in the box below, draw two climagraphs for the country depicted in Map 2-13.**

The country shown on this map has a cold, dry climate in the south, where most precipitation falls in the winter months, and a warm, wet climate in the north, where rain falls mostly in the spring. *This country is south of the equator. Remember that seasons in the Northern and Southern hemispheres are reversed.* Draw a climagraph for each part of the country.

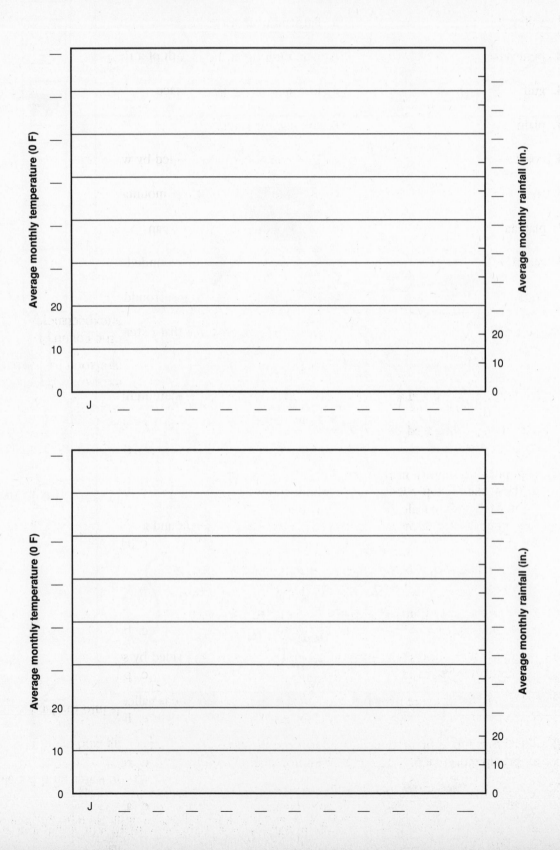

UNIT 2 REVIEW

A Match each term in Column A with its definition in Column B.

Column A   **Column B**

___f___ **1.** valley   a. flat land near sea level

___h___ **2.** peninsula   b. land built up at the mouth of a river

___j___ **3.** gulf   c. low spot in the land surface

___a___ **4.** plain   d. the highest landform

___g___ **5.** coast   e. land completely surrounded by water

___c___ **6.** basin   f. low land between hills or mountains

___i___ **7.** plateau   g. land beside a sea or an ocean

___e___ **8.** island   h. narrow piece of land surrounded by water on three sides

___b___ **9.** delta   i. flat land higher than the surrounding land

___d___ **10.** mountain   j. part of a body of water that extends into the land

B Write the letter of the word or words which will complete each statement correctly.

___b___ **1.** Elevation is measured from
   a. contours.   b. sea level.   c. mountain peaks.

___a___ **2.** A population density map shows
   a. how many people live   b. where resources are   c. where state capitals are located.
      in each square mile.   located.

___b___ **3.** Line graphs often show two kinds of information: an amount and a
   a. cause.   b. time.   c. effect.

___c___ **4.** A map which shows both temperature and rainfall is called a
   a. line graph.   b. circle graph.   c. climagraph.

___a___ **5.** When we wish to compare parts of a whole, we often use a
   a. circle graph.   b. line graph.   c. bar graph.

___a___ **6.** A graph which shows how the population of a country is divided by sex and age is called a
   a. population pyramid.   b. population pie chart.   c. population bar graph.

___c___ **7.** How long the average person in a country is expected to live is called the country's
   a. birth rate.   b. death rate.   c. life expectancy.

___c___ **8.** The things found or produced in an area can be shown on a
   a. precipitation map.   b. elevation map.   c. resource map.

___a___ **9.** Part of a body of water that extends into the land is called
   a. a bay.   b. an isthmus.   c. a delta.

_____b_____ **10.** The amount of elevation between contour lines on a contour map is called the
　　　　　a. contour measure.　　　　b. contour interval.　　　　c. interval measure.

_____c_____ **11.** Line graphs are useful because just a glance at one can tell you not only the trend, but also the
　　　　　a. time.　　　　　　　　b. percentage.　　　　　　　c. rate of change.

_____c_____ **12.** Another word for rainfall is
　　　　　a. temperature.　　　　　b. humidity.　　　　　　　c. precipitation.

_____c_____ **13.** A climagraph can look like two types of graphs put together. These two types of graphs are a line
　　　　　graph and a
　　　　　a. population pyramid.　　　b. circle graph.　　　　　c. bar graph.

_____a_____ **14.** Circle graphs are divided into
　　　　　a. 100 parts.　　　　　　b. 50 parts.　　　　　　　c. 150 parts.

_____c_____ **15.** Information presented in easy-to-read list form is called a
　　　　　a. pie graph.　　　　　　b. line graph.　　　　　　c. table.

C　**Answer these questions about the bar graph.**

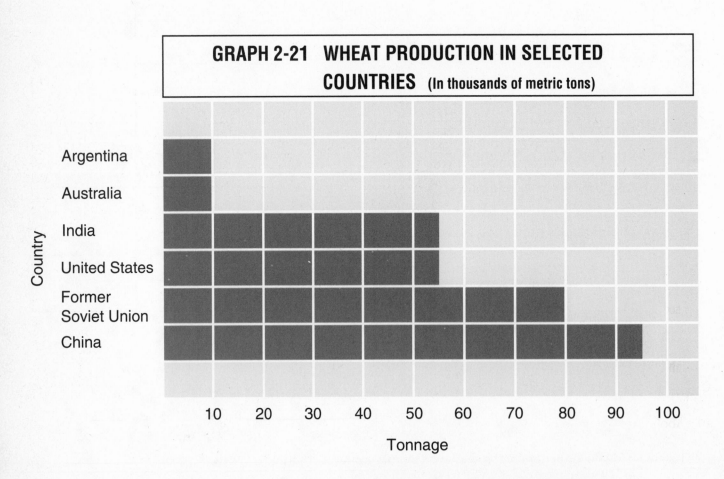

GRAPH 2-21　WHEAT PRODUCTION IN SELECTED COUNTRIES　(In thousands of metric tons)

1. About how much wheat was produced in the United States? _____54 thousand metric tons_____

2. Which countries produced the least wheat? The most? _____least—Argentina, Australia; most—China_____

3. About how much more wheat did China produce than India? _____40 metric tons_____

D **Answer these questions about the line graph.**

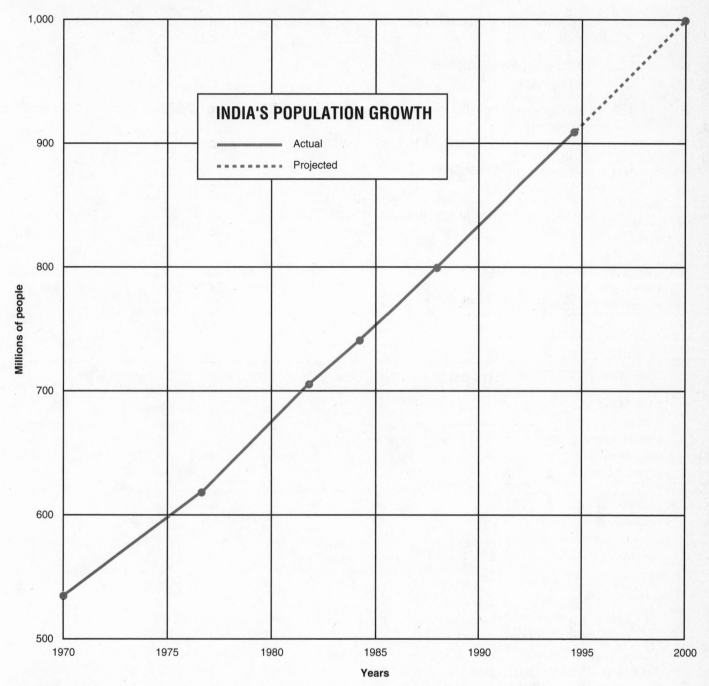

Graph 2-22

1. About how many people lived in India in 1994?   about 911 million

2. How would you desrcibe the trend shown by this graph?   The trend is up.

3. About how many people are expected to be added to the population of India between 1994 and 2000?

about 100 million

4. About how many people lived in India in 1970?   about 540 million

UNIT 3 Human-Environmental Relations

Humans have a "love-hate" relationship with their environment. We recognize how important our environment is. At the same time, we use it far beyond its capacity to clean itself. The result is what we call pollution. Pollution can affect the air we breathe, the water we drink, and the soil in which we grow our foods.

Not all dangers in the environment are made by humans. Volcanoes pour dirt into the air. Storms destroy homes and crops. Earthquakes rattle dishes and have made whole buildings collapse.

Human-environmental relations is one of the five themes of geography. People may change their environment, or they may be changed by it. For example, hundreds of years ago, Native Americans built farming villages in the American Southwest.

When the climate changed and became drier, Native Americans left their villages. Later, Hispanic and Anglo settlers set up mines, farms, and ranches in the same area. These later settlers brought water into the area for their needs. Rather than change themselves, they changed their environment.

A more recent example of how humans interact with their environment will be played out in your lifetime. For the last two hundred years, humans have been burning coal and oil at faster and faster rates. This burning has released a gas called carbon dioxide into the air. Carbon dioxide traps the heat of the sun close to the earth. The earth is slowly warming up. This warming could melt the ice at the North and South poles and flood many cities around the world.

Pollution Facts

1. Sitting quietly and breathing normally, the average person breathes about 16,000 quarts of air a day.
2. On the average, each person in the United States uses over 70 gallons of water a day in the home.
3. The average American family buys about 1 ton (2,000 pounds) of food every year and throws away about 10% of it.

Tourist Problem at the Bottom of the World

For years researchers have liked to use the continent of Antarctica as a place to do research. For instance some researchers conduct experiments to look at the results of isolation on humans. After all, when you are down near the South Pole, you can't count on too much company. Or can you?

It seems that Antarctica is quickly becoming a hot tourist spot. Thanks to cruise ships and charter airline services, more than 3,500 people a year are visiting the frozen below. They come to see the seal rookeries, the penguins and even the South Pole, itself. Scientists are now working to control the stream of tourists so that people can still visit Antarctica and the experiments can be conducted in peace. No one has asked the penguins and the seals how they feel about their visitors from the north.

The warming may also change the earth's weather, possibly causing crop failures in many areas. No one yet knows what changes will actually take place.

In this unit, you will study some causes and solutions to the problems of air and water pollution. However, you will learn that the relationship between humans and their environment is very complicated. We do not have nearly as many answers as we have questions.

Lesson 1 The Water Cycle and Pollution

What do you have in common with King Tut and dinosaurs? The rain that falls on you today may once have fallen on them.

How is this possible? The answer lies in what we call the **water cycle,** the movement of water from oceans, to air, to land, and back to the oceans. All the water on earth is used over and over again. As water moves through the water cycle, it becomes mixed with water from all over the earth—and sooner or later water that once fell on a dinosaur falls on you.

The Water Cycle

Illustration 3–1 shows how the water cycle works. The water cycle really has no beginning or ending point. It is an endless loop. However, let's begin with the water in the oceans. As the sun beats down on the oceans, it heats the water. Some of the water is changed into a gas by **evaporation.** The gas, called water vapor, rises into the air. The water vapor forms clouds, which drop rain or other forms of precipitation on the earth. Some of the rain is taken up by plants, which "breathe" the water vapor back into the air through their leaves. This is called **transpiration.** Some of the rain sinks into the ground (becomes **groundwater**) and eventually is pumped from wells or flows from springs. **Surface runoff** flows into rivers, which return it to the ocean. Then the cycle starts all over again.

Human Effects on the Water Cycle

Let's look at how humans affect the water cycle. Humans create a great deal of **pollution.** Pollution is something unclean in the **environment,** the world in which we live. We dump sewage into lakes and rivers. Factories billow smoke into the air and gush harmful chemicals into our rivers. Other wastes from factories are pumped into wells drilled deep into the ground.

Look again at the diagram of the water cycle. Find the part of the diagram labeled "groundwater." Suppose you own a factory which has a harmful chemical left over after goods are made. Your factory makes thousands of tons of this chemical each year. You have no place to store it. You must get rid of it. Many factories face this problem. The answer that many turn to is called **deepwell injection.** Wells are drilled deep into the ground, far deeper than groundwater can go. Dynamite is exploded at the bottom of the well to break up the rock. Then harmful wastes are pumped into the well. Over half the dangerous wastes produced in the United States are pumped down deepwells. Some of these wastes leak into the groundwater. Once the dangerous wastes enter the water cycle, it may be years before they are removed—if ever.

Surface runoff provides another way for wastes to enter the water cycle. Rain or snow falling on the ground picks up pollution from streets and garbage dumps and carries it into streams. Chemicals that

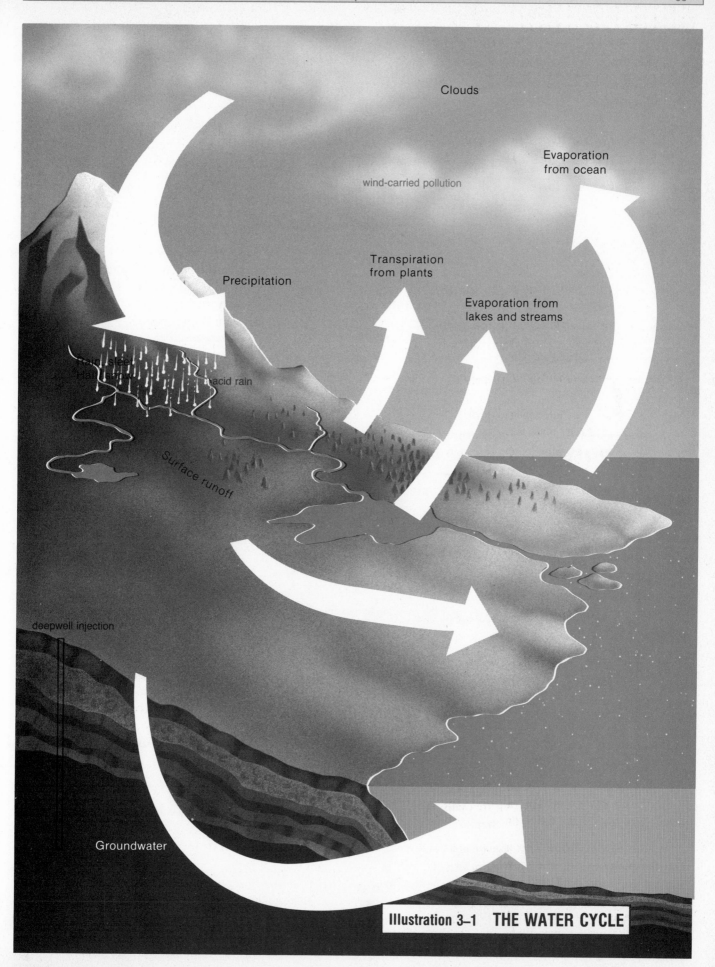

Clouds

Evaporation
from ocean

wind-carried pollution

Transpiration
from plants

Precipitation

Evaporation from
lakes and streams

Rain, sleet
Hail, snow

acid rain

Surface runoff

deepwell injection

Groundwater

Illustration 3–1   **THE WATER CYCLE**

farmers use to kill weeds and insects become part of surface runoff.

Wastes also enter the water cycle through the air. Each year in the United States alone, billions of pounds of poisons enter the air from factories, trash burning, garbage dumps, cars, and agricultural spraying. Some of these poisons fall directly into lakes and streams. Others are washed out of the air by rain or snow. In either case the poisons enter the water cycle.

Acid Rain

Germany is a good example of a country which has suffered a great deal of damage from **acid rain.** Acid rain is rain or snow that carries pollution. Most of the pollution in acid rain comes from the burning of fuels such as gasoline, coal, and oil. An acid created in the air falls with rain. Thousands of acres of trees in Germany have been killed by acid rain.

Acid rain is also a problem in the United States and Canada. All the fish have been killed in hundreds of lakes. Forests are filled with dead or dying trees. Where are the poisons doing this terrible damage coming from? They are coming from the exhausts of the cars, buses, and planes we use for travel. They are coming from the smokestacks of the factories that make the dozens of things we buy to make life more pleasant. They are coming from the smokestacks of the power plants that make the electricity that does most of the work in our homes. In order to make our lives more pleasant, we are poisoning the earth on which we live.

A number of years ago a seventh-grade student working on a class project in Austin, Texas, came up with a slogan: "You're the solution to water pollution." The Texas Department of Parks and Wildlife used that slogan to make people realize this important fact: People cause pollution, and only people can stop it.

Using Your Skills

A Write the meaning of each term.

1. water cycle   The water cycle is the movement of water from the ocean, to the air, to the land, and back to the ocean.

2. pollution   Pollution is something unclean in the environment.

3. deepwell injection   Deepwell injection is the pumping of harmful wastes into deep wells to get rid of them.

4. acid rain   Acid rain is rain or snow that carries pollution.

B Answer these questions.

1. Discuss the ways in which pollution can enter the water cycle.   Harmful wastes can be dumped into rivers, lakes, or the oceans. Runoff can carry poisons into the water. Wastes put into deep wells can leak into groundwater. Pollution in the air can fall into lakes or rivers, or it can be washed out by rainfall.

2. What are the ways in which people are causing pollution?   People cause pollution by dumping harmful wastes into the water. The factories that make things we use cause pollution. Our cars cause pollution.

3. On the diagram of the water cycle on page 95, draw and label pictures to show how pollution enters the water cycle. Include deepwell injection, wind-carried pollution, and acid rain.   Answers will vary.

Lesson 2   Understanding Causes of Air Pollution

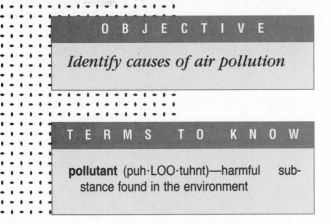

A volcano erupts in Mexico, spewing smoke and ash thousands of feet into the air. A bus drops off a load of children at school. A carpenter builds shelves for a new kitchen. A power plant makes electricity for a city miles away.

What do all these things have in common? As unlikely as it may seem, they all play a part in air pollution.

Causes of Air Pollution

When a volcano erupts, it can throw out huge clouds of dust. The dust can spread around the entire world. This dust can block some of the light of the sun and cause winters to be colder than normal for several years. When Mt. St. Helens erupted in the United States in 1980, the engines of many cars were damaged by dust they sucked in.

Air pollution can be caused by nature in other ways. Forest fires started by lightning can pollute the air with smoke. Windstorms can throw large amounts of dust into the air.

As powerful as nature is, however, people cause far more air pollution. This pollution is of two types: outdoor pollution and indoor pollution.

Outdoor air pollution comes from a variety of sources. Farmers cause a great deal of air pollution. As they plow fields, dust is thrown into the air. Winds blowing across fields left bare after crops are harvested pick up still more dust. Another source of pollution on farms comes from the use of chemicals to kill weeds and insects. Many of these chemicals are sprayed by planes or tractors. The wind can carry tiny drops far from the fields.

Dangers of Air Pollution

Factories, power plants, cars, planes, ships, and even lawn mowers also cause air pollution. In fact, anything which burns a fuel such as gasoline, oil, or natural gas adds to air pollution. The gases from burning fuels can cause eye and lung irritation. Air pollution from gasoline engines can reach such dangerous levels in large cities that children and people who have trouble breathing are asked to stay indoors. Some people can die from the effects of air pollution.

Staying indoors may not get a person away from all air pollution, however. Indoor air pollution is now known to be a serious health threat in the United States. In fact, the air in some buildings is worse than the air outdoors.

Indoor air pollution comes from many things. The table below shows the most common indoor **pollutants.** Pollutants are harmful substances found in the environment.

INDOOR POLLUTANTS

Pollutant	Where It Comes From	What It Does
Radon gas	rocks and soil	may cause lung cancer
Tobacco smoke	cigarettes, pipes, cigars	causes lung cancer and other diseases of the lungs
Asbestos	pipe insulation; ceiling and floor tiles	causes lung disease; cancer
Fungi, bacteria	air conditioners	cause allergies, asthma
Carbon monoxide	stoves and heating systems	cause headaches
Formaldehyde	plywood, particle board, foam insulation	causes eye, skin, and lung irritation; may cause cancer
Benzene	certain types of cleaners	may cause leukemia
Styrene	carpets, plastics	causes liver and kidney damage

Indoor pollution is a greater problem now than it was a few years ago. One reason is that houses today are being built to hold air in better. Windows and doors fit tighter. Pollutants such as cigarette smoke have a harder time getting out. Another reason indoor pollution is such a problem is that people spend up to 80 percent of their time indoors. Even low levels of pollution add up.

Air pollution from radon is largely a problem of air being trapped inside houses. Radon is a gas. You cannot see it or smell it. The soil and rock in many parts of the world contain radon. The gas can enter a house through the floor and walls. Radon from water wells comes into the house with water. If air can enter and leave the house easily, the radon escapes. However, a house with tight-fitting doors and windows may trap the radon. People who breathe high levels of radon over a number of years are more likely to get lung cancer. Map 3–1 shows areas in the United States and Canada which have a possible radon risk.

Buildings not only trap indoor pollution. They also create it. Asbestos and formaldehyde are found in many building materials. When these substances get loose inside buildings, they can be dangerous. By law, asbestos may no longer be used in buildings in the United States. However, many older buildings have asbestos in them. People who breathe air with tiny bits of asbestos in it may some day get cancer. Formaldehyde, which is often used in plywood and other building materials, can cause people to feel ill.

Two other common indoor pollutants are tobacco smoke and gases from burning fuels. Even people who do not smoke are affected by tobacco smoke. Children especially may be harmed by breathing air filled with tobacco smoke. Gases from fireplaces, stoves, and heaters can be harmful if they cannot escape from the house. In some cases enough gases can collect to cause death.

Using Your Skills

A   Answer these questions about air pollution.

1. Describe the two kinds of air pollution.   Outdoor air pollution can be caused by people or by nature. It includes pollutants

such as dust, smoke, and chemicals. Indoor air pollution is mostly caused by people. It includes pollutants such as tobacco smoke,

asbestos, formaldehyde, and gases given off by fires.

2. What are three things in nature that can cause air pollution?   volcanoes, wind storms, and forest fires

3. Describe two ways in which people cause outdoor air pollution.   Farmers cause air pollution by throwing dust and

chemicals into the air. Engines from cars, planes, and other things with motors also cause air pollution.

4. Why has indoor pollution become a problem?   Houses are built tighter than they used to be. Pollutants in the air cannot

get out as easily. Also, people spend most of their time indoors. They breathe the pollutants for many hours each day.

5. What kinds of health problems does the table in this lesson tell you can come from indoor air pollution?

Health problems range from headaches to lung cancer and leukemia. Indoor air pollution can cause liver and kidney damage, allergies,

asthma, and eye and skin irritations.

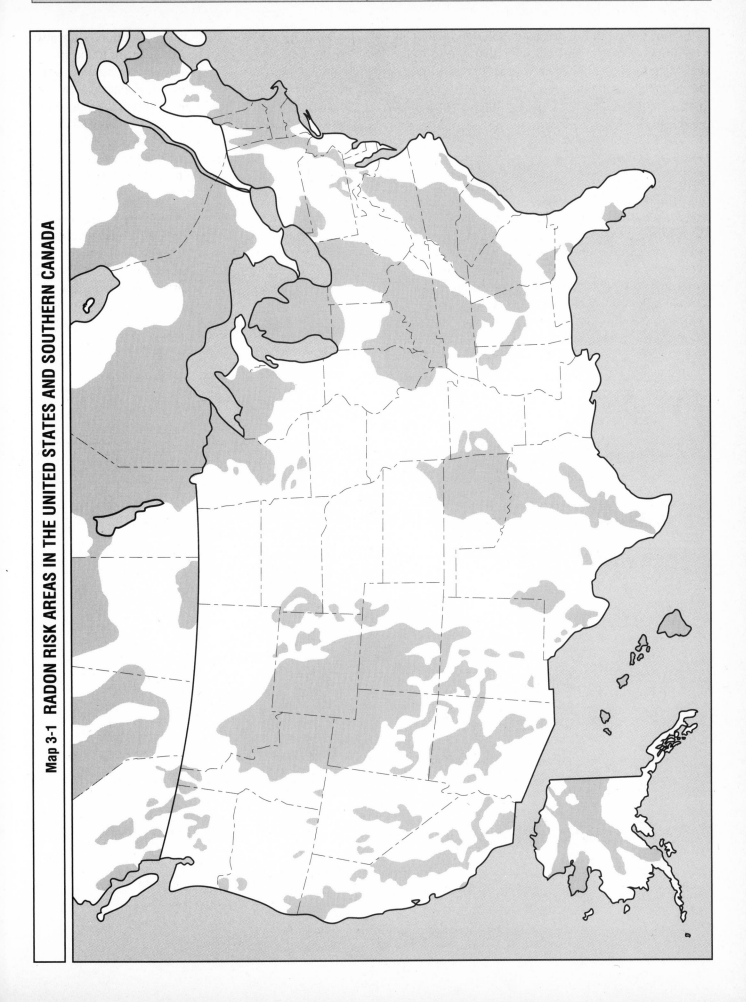

Map 3-1 RADON RISK AREAS IN THE UNITED STATES AND SOUTHERN CANADA

Lesson 3 Conserving Resources by Recycling

Do you like to pick up a hamburger, french fries, and a soft drink on your way to the park? When there is a good movie on TV, do you have a pizza delivered? Does your family ever eat frozen dinners? Do you ever wonder what happens to all the food and drink containers after you put them in the garbage?

Americans produce more garbage per person than any other people on earth. Each American throws away about five pounds of waste each day—three-fourths of a ton in a year. All Americans together produce enough trash in one year to fill 11 million garbage trucks. That's enough garbage trucks to make a traffic jam that would stretch around the world twice.

Getting Rid of Garbage

More than 90 percent of this trash is buried in giant holes in the ground called **landfills.** The largest landfill in the world is on Staten Island, New York. Every day 18,000 tons of trash from New York City are dumped there. This landfill will run out of room by the year 2000. No one knows what will happen to the trash then.

Over 9,000 landfills dot the American landscape. However, these landfills are also rapidly filling up with garbage. Over half the nation's landfills will be filled by the year 2000. Cities all over America are running out of places to put their garbage.

Garbage has always been something people would rather forget. "Out of sight, out of mind," the old saying goes, and people have been happy to have their garbage disappear from the curb twice a week as if by magic. But the closing of landfills has caused people around the country and around the world to take a new look at garbage.

Getting rid of garbage is a problem, and no one expects the problem to go away. However, progress is being made in the battle to keep our planet from being buried under its own refuse. The growing mountains of garbage are being fought in two main ways.

Burning Garbage

Garbage is being burned so that it will take up less room when buried. Over 100 plants in the United States burn garbage. Over 200 more are being planned. Some of these plants make steam to power factories, or make electricity to run motors. The ash left after the garbage is burned must still be buried. However, this takes up only about half as much space as the unburned garbage would have. One problem with this way of getting rid of garbage is that the burning causes air pollution. Things made of plastic may give off dangerous gases when burned.

Recycling Garbage

Garbage is being saved and put to new uses. There is an old saying that "One person's trash is another person's treasure." Many things that are thrown away can be saved and put to a new use. We call this **recycling.** Recycling saves money in several ways. Less room is needed in landfills. Valuable metals and other resources are saved. Often, recycling takes less energy than making new products.

Look at Graph 3–1. It shows what our garbage is made of. Paper, glass, metals, and even some plastics can be saved and made into new things.

Recycling is growing in the United States. However, there is much room for improvement. Even the best recycling programs in the United States save only about 20 percent of the garbage. By comparison, Japan recycles 95 percent of its bottles, 75 percent of its aluminum cans, and 50 percent of many other things. As a result, each person in Japan produces only half as much garbage as each person in the United States.

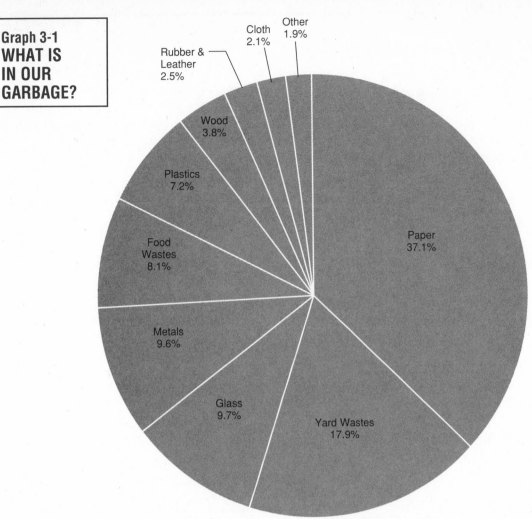

Graph 3-1
WHAT IS IN OUR GARBAGE?

Other 1.9%
Cloth 2.1%
Rubber & Leather 2.5%
Wood 3.8%
Plastics 7.2%
Food Wastes 8.1%
Metals 9.6%
Glass 9.7%
Yard Wastes 17.9%
Paper 37.1%

In some recycling, all garbage is dumped into giant machines that separate the metal and glass from the other garbage. The metal and glass are sold to companies that use them to make new things.

The best recycling systems have people separate their garbage in their homes rather than dumping it all into one can or bag. Usually each color of glass must be kept separate. "Tin" cans must be separated from aluminum cans. Plastic bottles go into another stack. Separating the garbage in the kitchen keeps costs down.

There are two reasons recycling is not used more widely. One reason is that many people do not want to go to the trouble to separate their garbage. Therefore, even the best programs get only about one-fourth of the people to take part. The second reason is that city and state governments have been slow to set up recycling programs. But as the garbage piles higher and the gates to more landfills are closed forever, more cities are starting to recycle. New Jersey, Rhode Island, and Connecticut have passed laws which require people to recycle. Massachusetts and New Jersey have set up large plants to recycle garbage.

How You Can Help

Many experts feel that the only way to solve the garbage problem is to get each citizen involved. How can you help? Here are some ways that work.

1. Don't buy things in containers that are made to be thrown away. Buy products in glass bottles or metal cans. Avoid plastic and paper containers whenever possible. Don't use grocery bags unless you really need them.

2. Recycle paper, metal, glass, and even plastic if at all possible. School paper drives are one way to recycle. Some cities have recycling centers where you can take glass, metal, paper, used motor oil, etc.

3. If there is no recycling program in your town, work to start one. City leaders know it often costs more money to bury trash than it does to recycle. If enough people are interested in recycling, a program may be started.

There are many ways to deal with the garbage problem. But everyone agrees this problem won't go away. And the longer the garbage problem is with us, the more it will smell.

Using Your Skills

A   Use the reading above to answer these questions.

1. What problem are cities all over America having with their garbage?   Landfills are getting full, and

cities are running out of room for their garbage.

2. What are two ways people are working to solve the garbage problem?   Some cities are burning garbage so it takes

up less room in landfills. Other cities are recycling garbage.

3. How does recycling garbage help solve the garbage problem?   Recycling saves room in landfills.

4. Why has recycling not been widely used in the United States?   Many people do not want to separate their trash. And

many cities and states have been slow to set up recycling programs.

5. How can you help solve the garbage problem?   by not buying things in throwaway containers; by recycling; by starting a

recycling program where there is none

B   Use the table below to answer these questions.

World Garbage Output

Country	Garbage Produced Yearly	
	Total (in thousands of metric tons)	Per person (in pounds)
United States	160,000	1,547
Australia	10,000	1,498
Austria	1,560	458
Canada	12,600	1,157
West Germany	20,780	744
Japan	40,225	757
Portugal	1,500	334

1. Which of the countries produces the most garbage per year? The most garbage per person?   the United States

2. Packaging for prepared foods is responsible for a large part of the garbage in the United States. In which two

countries above would you expect to find few such foods? Why?   Portugal and Austria. These two countries produce

far fewer pounds of garbage per person per year than the United States does.

Lesson 4 Hazardous and Toxic Waste Disposal

OBJECTIVE

Recognize that the disposal of hazardous wastes is a serious global problem

TERMS TO KNOW

hazardous (HAZ·erd·uhs)—dangerous

toxic waste (TAHX·ik wayst)—things people throw away that can be harmful

waste reduction (wayst ree·DUHK·shuhn) —the making of fewer wastes

Do you know where the water you drink comes from? Of course, it probably comes out of a faucet. But where was that water before it became part of the water supply? Was it in a lake or river, or was it pumped from a well? If you live on a farm or ranch, the chances are 9 in 10 that the water came from a well. If you live in a city, the chances are 7 in 10. Overall, about half of all Americans drink water from wells.

Danger in the Well

Some Americans have had the shock of learning that the water from their well is no longer safe to drink. In San José, California, for example, a chemical used in the manufacture of computer chips leaked from storage tanks into the groundwater. The chemical, TCA, may cause cancer. The drinking water for 20,000 people was spoiled. More than $40 million has been spent to stop the waste from spreading.

The problem of **toxic wastes** has been growing for a number of years. Toxic wastes are things people throw away which can be harmful. Toxic wastes are poison. Toxic wastes can damage or kill living things—trees, fish, birds, people.

Almost all toxic wastes begin with people. You and your family may be harming the environment with toxic wastes. Here is a list of some of the **hazardous** (dangerous) wastes that go into our trash every day. (Even if the product has been used, a small amount will remain in the container.)

HOUSEHOLD CLEANERS
- drain openers, oven cleaners, wood and metal cleaners and polishes, toilet bowl cleaners

HOME CARE PRODUCTS
- paint thinners, strippers, and removers; glues, paints, stains, varnishes

CAR CARE PRODUCTS
- used motor oil, parts cleaners, antifreeze, waxes and cleaners, oil and fuel additives

LAWN AND GARDEN PRODUCTS
- herbicides, pesticides, wood preservatives

OTHER
- batteries, fingernail polish remover, pool chemicals

Some experts feel that our country's landfills are becoming more dangerous than the dumps made for hazardous wastes. Hazardous waste dumps are required by law to be lined to keep wastes from leaking out. City landfills are not lined. Problems have arisen around many landfills as rainfall carries harmful wastes into underground water supplies. In a number of cities across the United States, everyone must drink bottled water because the city water supply has been spoiled by garbage.

Industry is one of the major producers of toxic wastes. In 1983 the United States produced 71 billion gallons of hazardous wastes—320 gallons for every person in the country. That's enough toxic wastes to cover Washington, D.C., two feet deep. Chemicals from industry make up most of this amount.

Industries use the chemicals to make products that people want to buy. When an industry is finished with the chemicals, it must find a way to get rid of them. Many chemical wastes are pumped down deepwells. Sometimes the wells leak, letting the chemicals into the groundwater. Many chemical wastes are stored in tanks. These tanks can also leak, spilling chemicals into the air and water.

The Superfund Law

Many years ago people did not give much thought to toxic wastes. Wastes were dumped into unlined pits, where they soaked into the ground. In 1980 Congress passed a law to help clean up these waste sites. This law was called the Superfund law because it set up a fund of $1.6 billion to clean up toxic waste sites. At the time the law was passed, government officials said that the way Americans had been disposing of toxic wastes was "the most grievous error in judgment we as a nation have ever made." Toxic wastes were called "a ticking time bomb ready to go off."

Under the 1980 Superfund law, cleanup work was started at about 100 toxic waste sites. Only 13 were completely cleaned up. It became clear that the problem was much bigger than had been thought. In 1986 a new Superfund law was passed. This new law set up a fund of $9 billion to clean up toxic waste sites.

Map 3–2 shows how widespread the toxic waste problem is in the United States. Find your state on the map. How many toxic waste sites are in your state? How many sites in your state are on the list of sites to be cleaned up under the new Superfund law? Will this law solve the toxic waste problem in your state?

There are a number of problems that must be solved before toxic wastes can be cleaned up in this country. The cleanup will be very expensive—as much as $100 billion. Who should pay for the cost?

Another problem is what to do with the wastes once they are cleaned up. Where will they be stored? No one wants to live next to a toxic waste site. One idea is to take liquid wastes out to sea in ships and burn them far out in the ocean. But what if the wastes are spilled while being loaded on the ship? What if the ship sinks in a storm and the chemicals leak into the ocean? And if the chemicals are burned, what if the wind carries dangerous smoke to land?

Is Waste Reduction the Answer?

Many people believe that the final answer may lie in simply stopping the production of toxic wastes. Already some industries have found that by using different methods, they can stop making so many toxic wastes. This is called **waste reduction.** Other industries have found that they can recycle wastes. Some experts feel that in the long run, waste reduction and recycling, along with cleanup efforts already under way, can solve the toxic waste problem.

Map 3-2 SUPERFUND SITES

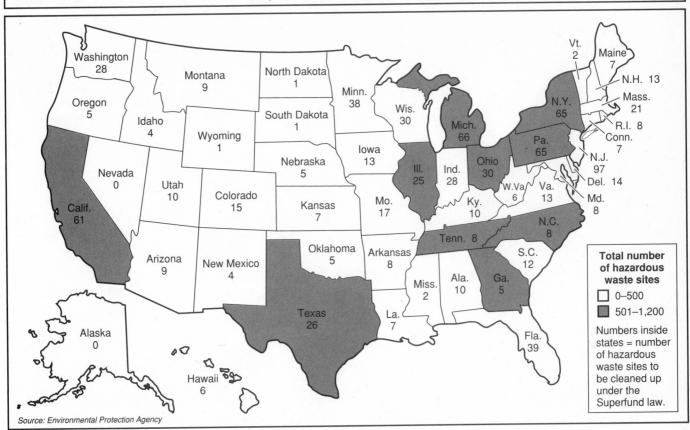

Source: Environmental Protection Agency

Using Your Skills

A   Answer these questions.

1. What are toxic wastes?   things that people throw away which can be harmful

2. How do you and your family add to the toxic waste problem?   by throwing such things as used motor oil, oven

cleaners, and batteries in the trash

3. What makes up most toxic wastes?   chemicals from industry

4. What is the Superfund supposed to do?   clean up toxic waste sites in the United States

5. Use Map 3.2 to answer this question. How many toxic waste sites are in New York? How many will be cleaned

up by the Superfund?   There are between 500 and 1,200 toxic waste sites in New York. Superfund will clean up 65.

6. What is waste reduction? How will it help solve the toxic waste problem?   Waste reduction is the use of different

methods by factories so that fewer toxic wastes are produced. Making fewer wastes will make it easier to dispose of toxic wastes

safely.

Lesson 5 Effects of Vertical Zonation

Do you live in the mountains? Or have you visited an area with mountains? Many people like the mountains because it is cooler there in the summer time. In fact, the higher you climb on a mountain, the cooler it will become. For every 1,000 feet you climb, the temperature will drop a little over 3°F. Visitors to Big Bend National Park in Texas can see this for themselves. The Chisos Mountains rise several thousand feet above the desert. Visitors may be sweating in 90°F in the desert. But by driving just two miles, they can climb into the mountains and enjoy temperatures in the 70's.

Look at Illustration 3–2. What is the temperature at the bottom of the mountain? What is the temperature at the top? What kind of clothing would you need to wear at the bottom of the mountain? What kind of clothing would you need at the top?

Effects of Altitude on Climate

The change in temperature due to altitude can affect the amount of rainfall an area gets. The cooler air gets, the less moisture it is able to hold. As warm moist air moves up a mountainside, it will be cooled. If the air becomes cool enough, rain will fall.

India provides a good example of what can happen. During the summer, winds blow across the Bay of Bengal toward India, picking up moisture from the sea. As the winds climb higher and higher, more and more rain falls. Dacca, India, in the lowlands near the sea, gets about 49 inches (124.5 centimeters) in four months. Only about 100 miles away, the village of Cherrapunji receives an average of 328 inches (833.1 centimeters) in the same four months. What makes the difference? Cherrapunji is about 4,000 feet higher than Dacca. As the air rises and cools, it drops its moisture on Cherrapunji.

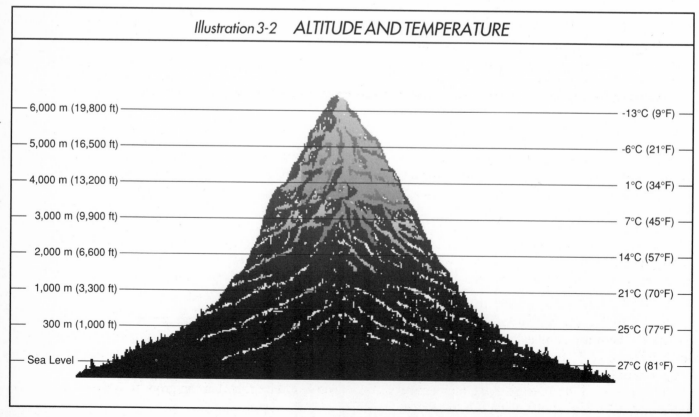

Illustration 3-2   ALTITUDE AND TEMPERATURE

Altitude	Temperature
6,000 m (19,800 ft)	-13°C (9°F)
5,000 m (16,500 ft)	-6°C (21°F)
4,000 m (13,200 ft)	1°C (34°F)
3,000 m (9,900 ft)	7°C (45°F)
2,000 m (6,600 ft)	14°C (57°F)
1,000 m (3,300 ft)	21°C (70°F)
300 m (1,000 ft)	25°C (77°F)
Sea Level	27°C (81°F)

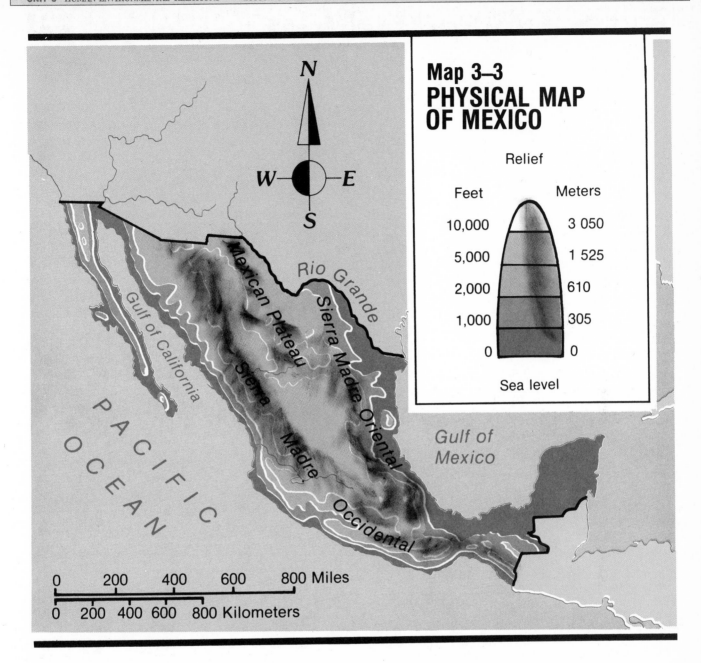

Map 3–3
PHYSICAL MAP OF MEXICO

Relief

Feet		Meters
10,000		3 050
5,000		1 525
2,000		610
1,000		305
0		0

Sea level

The change in climate due to altitude is called **vertical zonation.** Vertical zonation is especially important in countries near the equator. People prefer to live where the temperature is not too hot nor too cold. Look at Map 3–3. As you would expect, the climate is very hot in the lowlands along the coast. What happens to the elevation as you move away from the coast? Where would you expect most of the people of Mexico to live?

Mexico is often divided into three zones based on elevation. The warm zone is located at elevations from sea level to about 3,500 feet (1,000 meters). The temperature in the warm zone may reach over 90°F (33°C) for most of the year. Freezing temperatures are almost unknown. The cool zone is located at elevations from 3,500 feet to about 7,000 feet (2,000 meters). Daytime temperatures in the cool zone range from 65°F to 75°F (17°C to 22°C). Freezing temperatures at night are possible during the winter. Most of the people of Mexico live in the cool zone. Above 7,000 feet is the cold zone. Here daytime temperatures average between 55°F and 65°F (12°C to 17°C). Freezing temperatures often occur at night during the winter months.

Vertical zonation can affect where people choose to live. It also affects the kinds of crops they can grow. Different crops need different amounts of rainfall and temperatures to grow well. Illustration 3–3 shows some of the crops that can grow at different altitudes.

Illustration 3-3 VERTICAL ZONATION AND AGRICULTURE

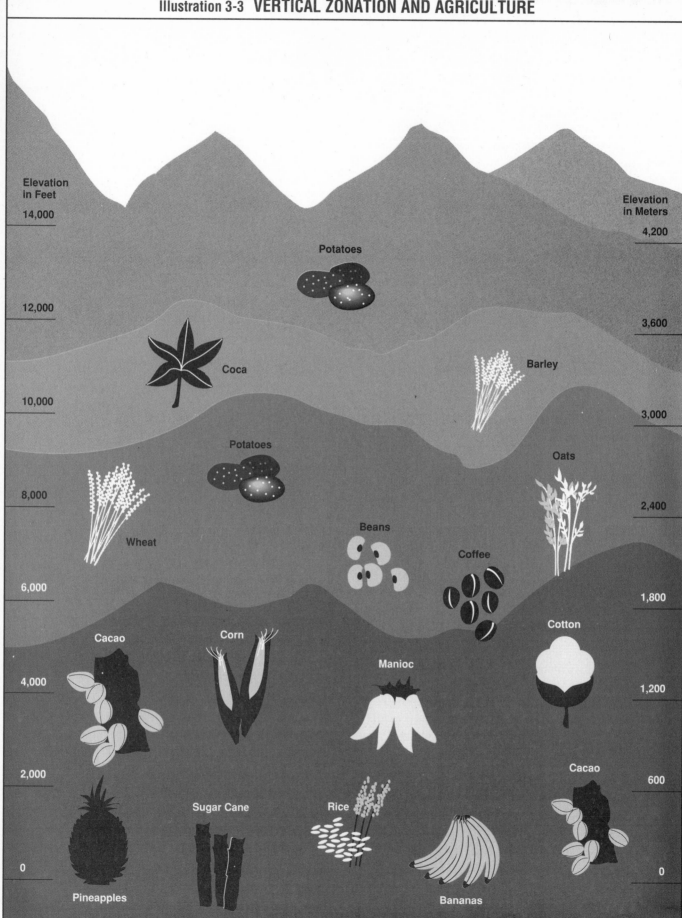

Using Your Skills

A   **Answer these questions.**

1. What is vertical zonation?   Vertical zonation is the change in climate due to elevation.

2. What happens to temperature as elevation increases?   Temperature goes down, or decreases.

3. In addition to temperature, what other part of climate can be affected by elevation?   rainfall

4. How does vertical zonation affect where people live?   People often choose to live at higher elevations, where

   temperatures are cooler.

5. How does vertical zonation affect agriculture?   The kinds of crops people can grow are affected by temperature and

   rainfall.

6. Use the illustration "Vertical Zonation and Agriculture" to answer these questions.

   a. What crops can be grown at elevations between sea level and 2,000 feet?   pineapples, sugarcane, rice,

      bananas, and cacao

   b. Which of these crops can also be grown between 4,000 and 6,000 feet?   cacao

   c. At about what elevation can coffee be grown?   between 6,000 and 8,000 feet

   d. What crop can be grown at the highest elevation?   potatoes

   e. What crops could be grown in Mexico's cool zone (between 3,500 and 7,000 feet)?   cacao, corn, manioc,

      cotton, wheat, beans, coffee, oats

7. What is the elevation where you live?   Answers will vary.

8. How does elevation affect the climate where you live?   Answers will vary.

9. How does elevation affect agriculture where you live?   Answers will vary.

OBJECTIVE

Be aware of danger to life and property from environmental hazards such as hurricanes, tornadoes, and lightning

TERMS TO KNOW

hurricane (HER·uh·kayn)—largest storm in nature

hurricane warning (HER·uh·kayn WAR·ning)—notice that a hurricane is expected to strike a particular location within 24 hours

hurricane watch (HER·uh·kayn wach)—notice that a hurricane is within 24 hours of striking somewhere

earthquake (ERTH·kwayk)—strong shaking of the earth

lightning (LYT·ning)—electricity passing between a cloud and the ground

tornado (tor·NAY·doh)—most violent storm in nature

The sound may be hard to hear at first over the pounding of hail on the roof, rain lashing at the windows, or thunder and lightning. People who have heard the sound say it is like the buzzing of 10 million bees. Others say it sounds like a train running through the living room, or a jet plane landing on the roof. Once you have heard it, you will never forget it. It is the sound of a tornado about to strike.

Tornado Safety

A **tornado** is the most violent storm in nature. It can destroy anything in its path. Tornadoes can happen anywhere in the United States. However, they are most common in the United States east of the Rocky Mountains. The peak months for tornadoes are April, May, and June, but they can occur in any month. Map 3–4 shows the months of peak tornado activity in the United States.

The only safe place in a tornado is a strong underground shelter. Second best is the corner of a basement *toward the tornado.* Most injuries in tornadoes are from flying objects such as glass and lumber. If you are in the corner toward the tornado, most objects will be blown over and away from you. If the building has no basement, go to a room, hallway, or closet *in the center of the house.* Get under a piece of furniture or pull a mattress over you.

Map 3-4 CONTINENTAL U.S. PEAK TORNADO ACTIVITY

People once thought that houses "exploded" in tornadoes. It was thought that opening windows would keep the air pressure equal inside and outside and save the house. *Opening windows is no longer recommended.* We now know that tornadoes simply blow houses down.

If you are in a car outside a city, you may be able to outrun a tornado. Drive at right angles to the direction the tornado is traveling. However, it is still best to take shelter. If you are in a car in a city, *take shelter.* If many people try to outrun a tornado, a traffic jam can trap you in its path. A car is one of the most dangerous places you can be during a tornado.

Outdoors, look for a place to hide. If there is no strong building around, hide in a ditch.

Lightning Safety

Tornadoes are usually born in strong thunderstorms. Most thunderstorms do not make tornadoes, but all have **lightning.** Lightning is electricity passing between the ground and a cloud. It is very dangerous. On average, more people are killed by lightning each year in the United States than by any other weather event. Lightning also starts many forest fires each year.

The best protection against lightning is to stay indoors during thunderstorms. Stay away from open windows, metal water pipes, stoves and refrigerators, and telephones. If you are caught outdoors and cannot get to a building, the safest place is in a car with windows rolled up. If you are on foot, stay away from high ground, wire fences, or trees that stand by themselves. If you are on a tractor, get off. Lightning often strikes the tallest object—which could be you sitting atop the tractor.

Sometimes you will have a few seconds warning before lightning strikes. If your hair begins to stand on end, *squat as low as possible.* You are about to be struck by lightning. DO NOT LIE FLAT ON THE GROUND. If you lie flat and lightning strikes the ground near you, it can travel through your body from end to end, passing through your heart and killing you. If you squat, the electricity will probably not pass through your body.

Hurricane Safety

The largest storms in nature are **hurricanes.** Hurricanes form over the ocean and bring strong winds and heavy rains. Hurricanes can push ocean water onto the land. The heavy rains can cause rivers to rise. Many deaths from hurricanes are caused by flooding. Tornadoes may also come with hurricanes.

Hurricanes are most dangerous to cities along the seacoast. When a hurricane is within 24 hours of striking land somewhere along the coast, a **hurricane watch** will be issued. A **hurricane warning** means that a storm is expected to strike a particular part of the coast within 24 hours.

The safest place in a hurricane is somewhere else. When a hurricane warning is issued, *leave.* Driving a hundred miles or more inland may save your life. Many people try to "ride out" the storm in order to "protect" their property. There is nothing anyone can do to protect property in a major storm.

Protection from Freezing

Cold is another danger that takes many lives each year. However, it is not the cold itself that kills. It is the loss of body heat. A person caught outside in wet clothing and in high winds can freeze to death at a temperature of 50°F (10°C).

Being prepared is the best safeguard against cold. If you will be outside in cold weather, dress warmly. Several layers of lighter clothing will keep you warmer than one or two very heavy layers. Mittens will keep hands warmer than gloves. Several pairs of socks will help protect your feet.

Many people are trapped in their cars by winter storms. If you must travel in snow or ice storms, carry food and extra clothing with you. Have blankets, sleeping bags, candles, matches, a shovel, and a tow rope with you. If you are trapped by snow, you can run the motor for a few minutes at a time to keep warm. However, you *must* keep snow shoveled away from the car. Otherwise the exhaust gases can enter the car and kill you.

If part of your body begins to freeze, it will first feel very cold. Next you will have a burning feeling, then no feeling. If you can, warm the body part as quickly as possible. If possible, put the part in warm water. If this is not possible, hold it next to your body. DO NOT RUB THE FROZEN PART WITH SNOW OR ICE.

Earthquake Safety

Earthquakes are a strong shaking of the earth. Most people are killed in earthquakes by falling buildings. If you are inside when an earthquake strikes, stand in a doorway. The strong doorway may protect you. You can also seek refuge under a desk or other sturdy piece of furniture. Do not run outside. Many people are injured or killed by falling debris such as trees or telephone poles while trying to run outside. If you are outside, go to an open area with no tall buildings nearby.

Using Your Skills

A Use your own paper to make a list of things you should do to protect yourself against tornadoes, hurricanes, lightning, cold, and earthquakes. Discuss these things with your family. Make a family plan for what to do when bad weather threatens. Answers will vary. Lists should stress safety tips given in the reading. Go over students' lists with them to be sure they understand what to do in each event. Stress to students that the important thing about this lesson is not knowing how to answer questions, but knowing what to do if weather threatens their lives.

B Use the table below to answer the questions.

Wind Chill Table

Temperature (°F)	Wind Speed (miles per hour)								
	5	10	15	20	25	30	35	40	45
	Wind Chill (Equivalent Temperature)								
35°	33°	22°	16°	12°	8°	6°	4°	3°	2°
30°	27°	16°	9°	4°	1°	−2°	−4°	−5°	−6°
25°	21°	10°	2°	−3°	−7°	−10°	−12°	−13°	−14°
20°	16°	3°	−5°	−10°	−15°	−18°	−20°	−21°	−22°
15°	12°	−3°	−11°	−17°	−22°	−25°	−27°	−29°	−30°
10°	7°	−9°	−18°	−24°	−29°	−33°	−35°	−37°	−38°
5°	0°	−15°	−25°	−31°	−36°	−41°	−43°	−45°	−46°
0°	−5°	−22°	−31°	−39°	−44°	−49°	−52°	−53°	−54°

Wind chill temperatures shown in **dark type** indicate a danger of freezing exposed flesh. Wind chill is what the temperature feels like if it is windy.

1. If the temperature is 15°F and the wind is blowing at 20 miles per hour, what is the wind chill? _____

   −17°F

2. You should be careful not to expose flesh at or below what wind chill temperature? How can you tell?

   −4°F. Temperatures shown in dark type indicate a danger of freezing exposed flesh. The highest wind chill temperature shown in dark type is −4°.

Lesson 7 Human Adaptation to Difficult Environments

OBJECTIVE

Describe ways in which humans adapt to different physical environments

TERMS TO KNOW

greenhouse effect (GREEN-hows ee-FEKT)—warming of the earth caused by gases that trap heat in the atmosphere

technology (tek-NAHL-uh-gee)—the use of tools and skills to make life easier

How would you like to live in a place where snow and ice cover the ground most of the year? Would you rather live in a place where rain almost never falls, and heat bakes the land year around? Most of us would rather live someplace more pleasant. However, some people do live in places which are very dry, or very hot, or very cold. How they live holds some lessons for the rest of us.

Adapting to the Environment

The Bushmen of southwestern Africa live in one of the world's driest deserts. They were pushed onto this land many years ago by other people who moved onto the better land to the east. The Bushmen have a hard life. There is little food in the desert. They have to eat almost anything, even bugs and frogs. They dig what roots they can find from under the ground and eat these, too. Often they go hungry. Because there is so little food, there are very few Bushmen.

One of the few tools Bushmen use is a pointed stick. They use the stick to dig roots from under the ground. Hunting is done with spears and the bow and arrow. Bushmen move about often, looking for new food supplies. Since they are always moving, they do not build houses. A few sticks in the ground, bent over and covered with grass, make a home for the few days it is needed.

When Bushmen move, they must carry everything themselves. They have no animals to help them.

Bushmen do not weigh themselves down with many belongings. One thing they must carry is water. They gather ostrich eggs, which are very large, and fill them with water.

After Bushmen have left an area, it would be very hard to tell that they had been there. Bushmen have made little use of **technology.** Technology is the use of tools and skills to make life easier. Instead of using technology to change their world, the Bushmen have learned to live with their world as it is. They have adapted to their environment. Because they use few tools, they make few changes in their environment. Because they have no cars, they have no roads. Because they have very few belongings, they have little to throw away. There are no Bushmen landfills. Because they have only simple tools, and no houses, they have no power plants to make electricity. The Bushmen lead very hard lives. But they do not have major problems with air and water pollution. Nor do they create many such problems.

The Eskimos are another example of a people who have learned to live in a difficult environment. They live in the far north of North America and Asia. Snow and ice cover the land for most of the year.

For many years the Eskimos used very little technology. For example, they made their houses of the one thing they have too much of—snow. These snow houses, or igloos, were heated by lamps which burn oil from animals. The Eskimos wore fur clothing made from the skins of animals. They hunted with bows and arrows and spears. Most of their tools were made from animal bones or from pieces of wood they found washed up along the beach.

Eskimo ways are changing. Some still follow the old ways some of the time, but the Eskimos have been quick to adopt new technology. Bows and arrows have been replaced with guns. Skin boats and wooden paddles have given way to metal boats and motors. The Eskimo sled pulled by dogs is being replaced by the snowmobile. And home to an Eskimo today is more likely to be a modern house than an igloo.

These change have not come without problems. When Eskimos began to want guns and houses and snowmobiles, they had to have money to buy them. They made money by hunting and fishing. By using guns and modern boats, they were able to kill many more animals and catch more fish. Before long, there were not as many animals and fish. It became harder and harder to earn the money needed to pay for all the new things the Eskimos wanted. Today, the number of animals the Eskimos kill must be limited. Otherwise, all the animals might be killed off.

Using Technology to Change the Environment

Think about how other people in the world have dealt with the same kinds of problems that faced the Eskimos and the Bushmen. Where it is too hot, people use air conditioning. Where it is too cold, people burn wood, gas, or oil or use electricity to heat their homes. If there is not enough food in an area to feed all the people who live there, food is brought in from far away. Factories turn out goods to feed, clothe, and amuse us. All kinds of work have been made easier by the use of machines.

Few of us would want to live as the Eskimos and Bushmen live. However, the pleasant lives we lead do carry a price tag. We live in a world which makes a great use of technology. Technology lets us feed, clothe, and house more people. It makes life more

pleasant. But there is a catch. Technology solves many problems, but it creates new ones. The pollution of our air, water, and land has been caused by our use of technology.

One of the problems caused by the use of technology is called the **greenhouse effect.** The greenhouse effect is a slow warming of the earth caused by the burning of fuels such as coal, oil, and gas. This warming started when people began burning very large amounts of these fuels about a hundred years ago. Burning these fuels makes a gas called carbon dioxide. Carbon dioxide traps heat from the sun. This makes the earth warm slowly over time. At the present rate of warming, the earth could be seven or eight degrees warmer in about sixty years. This does not sound like much, but it might be enough to cause great changes in the earth's climate. The areas which now grow most of the world's food could become deserts. The ice covering the North and South poles could melt. The melting ice could raise the oceans enough to flood many of the world's great cities.

Using Your Skills

A   **Answer these questions.**

1. What is technology?   the use of tools and skills to make life easier

2. How has the use of technology changed our world?   The use of technology has made our lives easier. It has also

caused air, water, and land pollution.

3. What is the greenhouse effect?   a slow warming of the earth caused by the burning of coal, oil, and gas

4. How may the greenhouse effect change our world?   Land which now grows most of the earth's food may become

desert. Rising oceans may flood many of the world's cities.

B   **Complete the chart below, showing how your life would be different without the use of technology. Remember: without technology you could use only tools powered by your own muscles. Use the headings shown below.**

Life Without Technology

With Technology	Without Technology
Food	Answers will vary but should point out that life as we know it would not exist without technology.
Clothing	
Shelter	
Transportation	

O B J E C T I V E

Understand the problems associated with overpopulation in developing nations

T E R M S   T O   K N O W

birth rate (berth rayt)—how many people are born each year

death rate (deth rayt)—how many people die each year

developing nation (duh·VEL·uh·ping NAY·shuhn)—country in which most of the people still depend on agriculture for their living

Have you ever gone to a store to buy something only to find that none was left? Perhaps you were told to come back the next day, or next week. You may not have been pleased, but you probably did not suffer lasting harm. But what if you had been shopping for food? What if all the food was gone?

More and more, food shortages are becoming a problem in **developing nations.** Developing nations are countries in which most of the people still depend on agriculture for their living. Farming methods used in these countries are often very old-fashioned. Few modern machines are used. The use of fertilizers and bug killers to increase the amount of food grown is almost unknown. Most developing countries have trouble growing enough food to feed their people.

Why World Population Is Increasing

Population growth is another problem in developing countries. People in developing countries often tend to have large families. One reason is that parents depend on their children to take care of them when they are too old to work. A man and woman who have many children are more likely to be taken care of in their old age. What if a family had only two children, and both died? Who would take care of the parents? Having many children makes it more likely that some will live to care for the parents.

But population growth does not depend just on how many people are born each year, the **birthrate.** Population growth also depends on the **death rate,** how many people die each year. In the last 40 years, the death rate in almost every country in the world has fallen a great deal. This has happened mainly

because of better medical care. People have also had more and better food to eat in many countries.

While the death rate has fallen in every developing country in the world, birth rates have remained high in most of them. This has caused a huge increase in the world's population in just a few years. Most of this increase has come in the developing countries, the very ones least able to feed more people.

Look at Map 3–5. Each part of the world shown has two figures printed on it. The top figure shows the average number of children each family in that part of the world has. The bottom figure shows the value of goods and services produced by each person in that part of the world each year. The larger this figure is, the more money people in that country are likely to have to spend.

You can see that the average family in Europe has 1.6 children. Each person in Europe produces $11,990 worth of goods and services each year. What are the figures for Africa? For Mexico? For Latin America? For Asia? Which parts of the world have the most children per family? Which parts of the world have the least money to spend?

Effects of Rapidly Increasing Population

Part of the problem is that a large percentage of the population of developing countries is made up of children. Before better medical care, many children died at an early age. For example, in Africa in 1950, 182 children out of every 1,000 born died before reaching their first birthday. By 1980 only 125 out of 1,000 died. In South America the rate fell from 126 to 70. In Asia it dropped from 155 to 95. These figures are still far higher than in developed countries. In the United States, for example, only 14 of 1,000 children die before their first birthday.

The lowered death rate in developing countries has caused the number of young children to grow very quickly. This means that there will be many people to work and buy goods in future years. However, large numbers of children are a drain on a country's resources. The children need food, clothing, schools, doctors, and much more. Developing countries have to spend most of their money just feeding their people. Little money is left for better roads and schools. Little money is left for tractors to grow more food, or factories to make more jobs.

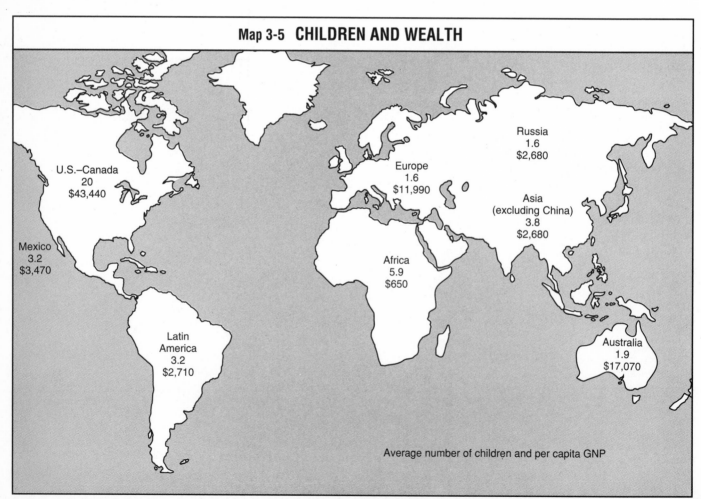

Map 3-5 CHILDREN AND WEALTH

Russia
1.6
$2,680

Europe
1.6
$11,990

Asia
(excluding China)
3.8
$2,680

U.S.–Canada
20
$43,440

Mexico
3.2
$3,470

Africa
5.9
$650

Latin
America
3.2
$2,710

Australia
1.9
$17,070

Average number of children and per capita GNP

Source: *World Population Data Sheet 1994*

Using Your Skills

A   **Answer these questions about Graph 3–2.**

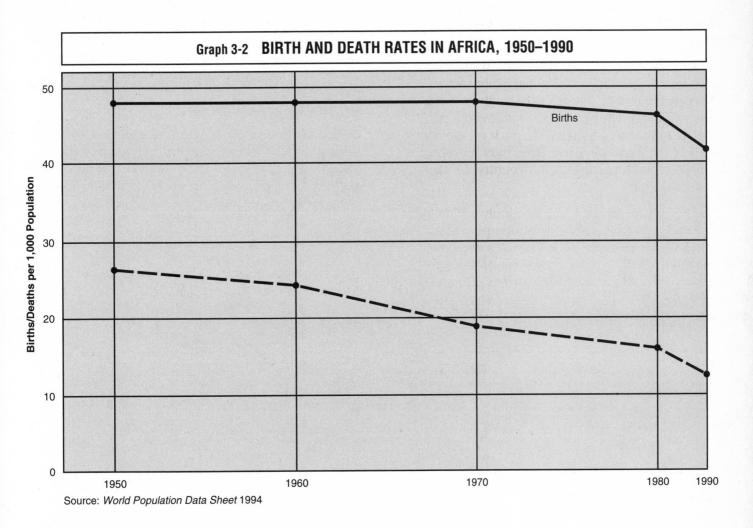

Graph 3-2   **BIRTH AND DEATH RATES IN AFRICA, 1950–1990**

Births/Deaths per 1,000 Population

Births

Source: *World Population Data Sheet* 1994

1. What is the title of this graph?   Birth and Death Rates in Africa, 1950–1990

2. What does the solid line on the graph stand for?   births per 1,000 population

3. What does the broken line on the graph stand for?   deaths per 1,000 population

4. What happened to the birth rate in Africa between 1950 and 1990?   It went down from 48 births per 1,000

population to 42 births per 1,000.

5. What happened to the death rate in Africa between 1950 and 1990?   It dropped from 27 deaths per 1,000

population to 13 deaths per 1,000.

6. What will happen to the population of Africa if the birth rate remains high and the death rate continues to fall?

The population will grow even faster.

B   Use Graph 3–3 to answer these questions.

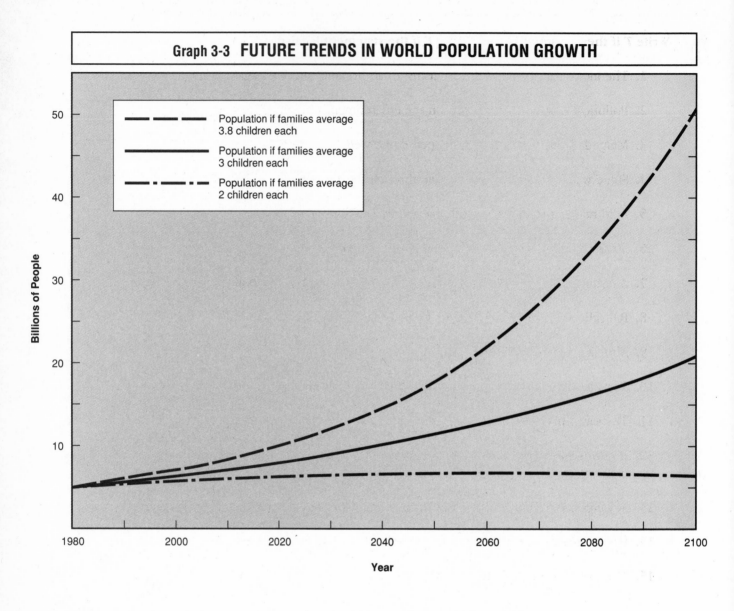

Graph 3-3   FUTURE TRENDS IN WORLD POPULATION GROWTH

Legend:
- Population if families average 3.8 children each
- Population if families average 3 children each
- Population if families average 2 children each

Y-axis: Billions of People

X-axis: Year

1. What does the title tell you this graph shows?   future trends in world population growth

2. The average world family today has 3.8 children. According to the graph, what will be the population of the world in the year 2100 if this continues?   about 50 billion people

3. What would happen to world population if each family had only two children?   The population would increase slowly until about the year 2030. After that it would stay about the same.

UNIT 3 REVIEW

A Write *T* if the statement is true. Write *F* if the statement is false.

F   1. The movement of water from oceans, to air, to land, and back again is called the waste cycle.

T   2. Pollution is something unclean in the environment.

T   3. Many dangerous wastes are pumped down wells.

T   4. Some wastes enter the water cycle through the air.

F   5. Acid rain is a form of pollution caused by landfills.

T   6. Most pollution is caused by the activities of people.

T   7. Anything which burns a fuel such as gasoline, oil, or gas adds to air pollution.

F   8. Recycling wastes causes landfills to fill up faster.

T   9. Almost all toxic wastes begin with people.

F   10. The change in climate due to time is called vertical zonation.

T   11. The safest place to be in a tornado is in an underground shelter.

F   12. If you think a tornado is coming, you should open the windows in your house, get in your car, and drive away.

F   13. More people are killed by floods in the United States each year than by any other weather.

F   14. Hurricanes strike quickly and give people no time to get to a safer place.

T   15. The use of technology to make our lives easier is a major cause of pollution.

B Write the letter of the word or words which will complete each statement correctly.

b   1. Water is changed into a gas by
   a. precipitation.   b. evaporation.   c. transportation.

c   2. Water which sinks into the ground is called
   a. sink water.   b. runoff.   c. groundwater.

b   3. Indoor air pollution is caused partly by
   a. too much air coming in.   b. too little air getting out.   c. smoke.

a   4. Each day the average American throws away how many pounds of waste?
   a. five pounds   b. three-fourths pound   c. two pounds

c   5. Over half the landfills in the United States will be filled by the year
   a. 2100.   b. 2150.   c. 2000.

b   6. The best recycling programs are ones in which
   a. a huge machine   b. people separate their   c. workers separate garbage at
      separates metal and   garbage.   recycling centers.
      glass from the other
      garbage.

_____a_____ **7.** The law which says that toxic wastes must be cleaned up is called the
   a. Superfund law.   b. Waste Cleanup Act.   c. Toxic Cleanup Act.

_____c_____ **8.** For every 1,000 feet you climb up a mountain, the temperature will
   a. rise 3°F.   b. drop 10°F.   c. drop 3°F.

_____a_____ **9.** The peak months for tornadoes are
   a. April, May, and June.   b. September, October, and   c. January and February.
                                November.

_____b_____ **10.** If you think you are about to be struck by lightning, it is safest to
   a. lie flat on the ground. b. squat as low as possible.   c. get under a tree.

C Use Map 3–6 on page 123 to complete the table.

Average Number of Tornadoes and Tornado Deaths in Selected States, 1953–1980

STATE	NO. OF TORNADOES	NO. OF DEATHS
California	3	0
Colorado	19	0
Texas	119	11
Illinois	27	5
Oklahoma	53	8
Tennessee	11	5
Maine	3	0
Florida	41	2
Ohio	14	5
Nevada	1	0

Map 3-6 AVERAGE NUMBER OF TORNADOS AND TORNADO DEATHS BY STATE, 1953–1980

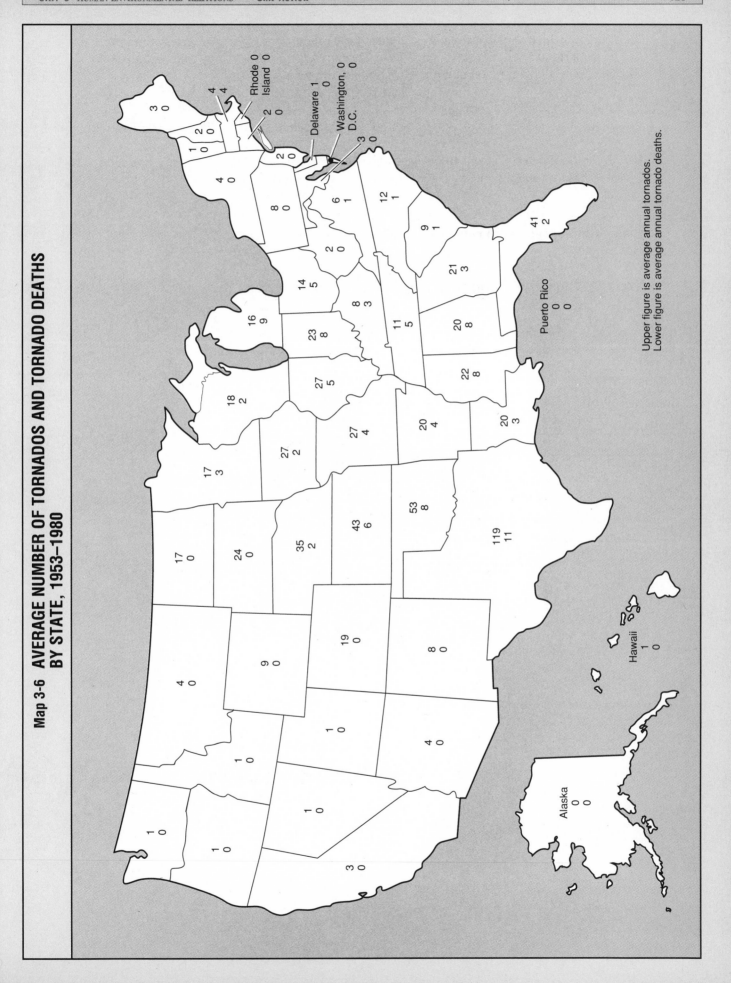

Rhode 0 Island 0
Delaware 1 0
Washington, 0 D.C. 0

3 0
4 4
1 0 2 0
2 0
4 0
8 0
6 1
12 1
9 1
41 2
2 0
21 3
14 5
8 3
11 5
20 8
16 9
23 8
22 8
18 2
27 5
20 4
20 3
17 3
27 2
27 4
53 8
17 0
24 0
35 2
43 6
119 11
4 0
9 0
19 0
8 0
1 0
4 0
1 0
1 0
1 0
1 0
3 0

Puerto Rico 0 0

Hawaii 1 0

Alaska 0 0

Upper figure is average annual tornados.
Lower figure is average annual tornado deaths.

OBJECTIVES

After completing this unit, you will be able to:

- demonstrate how humans interact within and among countries,

- discuss how peoples of the world are linked by trade and transportation,

- analyze how the United States is interdependent with other countries.

Every day each of us interacts with people around the world, even though we may not realize it. When you wake up in the morning and punch your pillow before going back to sleep, you may be touching cloth made from cotton grown in Egypt. Your breakfast may include bananas from Latin America. Your ride to school or work may take place in a car from Japan, Germany, or Korea. Even if the car is an American make, chances are that it was put together in Canada. Your jeans may have been sewn in Mexico.

Transportation—the movement of goods and people—makes our interactions with people in distant places possible. Imagine how different your life would be if the only goods you could buy were ones produced within your immediate area. Imagine how you would feel if you had to live out your life without ever traveling more than a few miles from where you were born.

Movement: humans interacting on the earth is one of the five basic themes of geography. We know that people are scattered unevenly across the earth. Yet we all interact with each other. We travel from place to place on business or for pleasure. We talk to people in other places. Television, books, and radio bring us ideas from other lands. Some of the clothes we wear, the foods we eat, and the gasoline that runs our cars come from other countries. In turn, we send other countries things we have that they need.

In short, we need other countries, and they need us. No country in the world today can survive strictly on its own. The word we use to describe this is

Fairy Tales from around the World

Growing up, all of us have listened to and read many fairy tales such as those collected and retold by the Brothers Grimm. Did you know that these fairy tales came from all over the world and were told and retold many times before anyone began writing them down?

The earliest records of these tales can help us to understand where they came from. "Sleeping Beauty" came from Italy, "Little Red Riding Hood" came from France, "Hansel and Gretel," and "Snow White and the Seven Dwarfs" were told in Germany, and "Goldilocks and the Three Bears" was first told in England. Over twelve hundred years ago, Chinese parents and grandparents first entertained their children by telling them the story we know as "Cinderella." Through communication, these stories have survived the years and spanned the globe to be told, read, watched, and enjoyed by children today.

What's So Great About a Wall?

Of all the structures that man has built on this Earth, only one can be viewed from the moon. This is the Great Wall of China. The Great Wall is over 4,000 miles long and historians believe that it might have taken almost 2,000 years to complete.

Originally built to help protect the Chinese people from Mongol raiders out of the North, it is now one of the world's greatest tourist attractions. When American astronauts first landed on the moon in 1969, they noticed that the great wall was the only man-made structure on earth that was visible to them.

interdependence. This is a big word that simply means that humans need each other.

Geography helps to explain the ways in which people, ideas, and goods move between countries. Geography helps to answer two important questions. The first question is: *Why* are things located in particular places? The second question is: *How* do those places influence our lives?

In this unit you will study some of the ways people interact with each other in the world today. You will learn how people in the United States interact with each other and with people around the world. You will see how a culture can spread across a large part of the earth. And you will gain a greater appreciation for just how complicated and interdependent our modern world is.

Lesson 1 Transportation in the United States

Do you know how much time you spend each day going from one place to another? If you were to keep track of this time each day for a week, the results might surprise you. Most of us spend many hours each week traveling. Moving people and goods from place to place is called **transportation.**

The United States has the best transportation system in the world. We use roads, railroads, waterways, airways, and pipelines to move ourselves and our products. The United States has over 4 million miles (6.4 million kilometers) of roads, streets, and highways. It has more miles of railroads than any other country. Seven of the ten busiest airports in the world are in the United States. Trucks carry 28 percent of America's goods. About 15 percent of goods are carried on water. And about 19 percent of goods shipped in the United States travel by pipeline. Railroads carry about 37 percent of goods.

This story can be repeated endless times. Many of the products you use each day are made in some other part of the country. The raw materials that are used to make these products come from many parts of the world. Transportation makes it possible to gather the raw materials and ship the finished goods to market. Transportation also makes it possible for workers to get to the factory to make the goods.

Transportation and Trade

Transportation is so important to the United States because our country depends on **trade.** Trade is the

buying and selling of goods. No part of the United States has everything it needs. People in the different parts of the country trade with each other for the things they need. Cotton from Mississippi may be sold to a factory in New York. The money may be used to buy a car made in Detroit. A car bought in this way may then be used for a vacation trip to California. During the trip the car may use gasoline made from oil shipped by pipeline from Texas or by water from Louisiana.

Transportation and Interdependence

Different parts of the United States depend on each other for goods they do not produce themselves. The United States also depends on other countries for some things. In turn, other countries depend on the United States. Depending on each other is called interdependence. Put simply, it means that we need others, and they need us.

One reason we *can* depend on each other is because transportation makes it possible. Before good roads, railroads, and airplanes were developed, people had to depend more on themselves and others nearby. Goods could be shipped, but travel was slow and expensive. Most people grew most of their own food and made most of their own clothing. Today shipping goods is fast and cheap. No matter where you live, your grocery store can have fruits from Texas, Florida, and California, meats from Iowa and Kansas, and fresh fish from the Atlantic and Pacific oceans. Fast, inexpensive transportation makes this possible.

Using Your Skills

A   **Write the meaning of each word.**

1. transportation   the moving of people and goods from one place to another

2. trade   the buying and selling of goods

3. interdependence   depending on each other

B   **Answer these questions.**

1. Why is transportation important to the United States?   Transportation is important to the United States because our

country depends on trade.

2. What part does transportation play in trade?   Transportation carries goods from one part of the country to another.

3. How does transportation make interdependence possible?   Fast, cheap transportation makes it possible for us to buy

goods from many different places.

4. Name a product that comes from another state or another country, such as a food, car, or appliance that you

used today, tell where it was originally grown or built, and tell how it was shipped to your city or town.

Answers will vary.

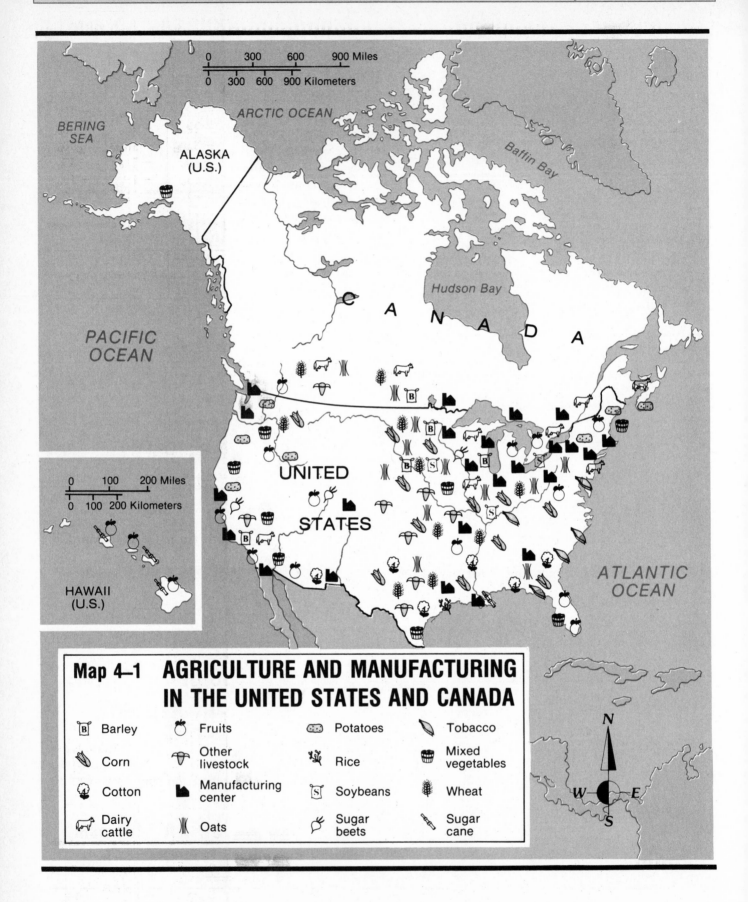

Map 4-1 AGRICULTURE AND MANUFACTURING IN THE UNITED STATES AND CANADA

C   **Use Maps 4–1 and 4–2 to answer the questions.**

1. What cities in the United States have both a major airport and a major seaport?   New York, New Orleans,

   Houston, Los Angeles, Seattle, Honolulu, and Chicago

2. Compare the two maps. What connection can you see between the amount of agriculture and manufacturing

   and the number of highways and railroads? Why do you think this is so?   There are more highways and railroads

   where there is more agriculture and manufacturing. This is because there are more people and goods to be moved.

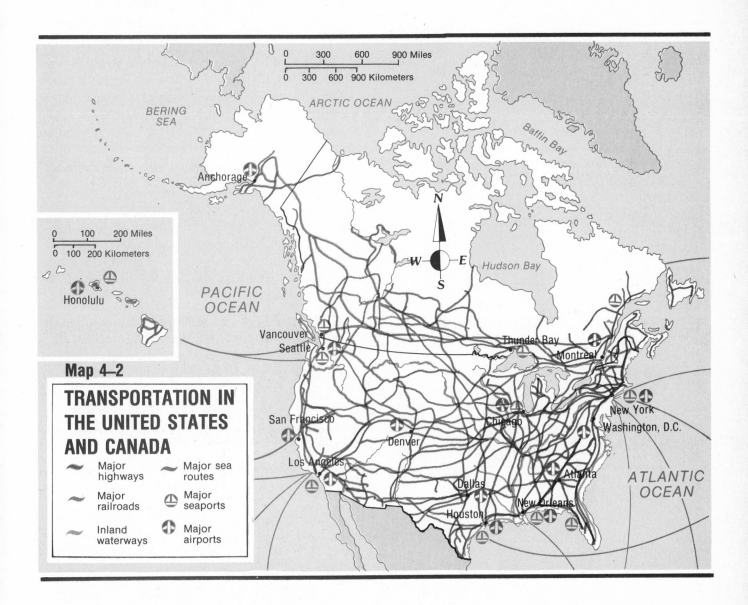

Map 4–2

TRANSPORTATION IN THE UNITED STATES AND CANADA

~ Major highways
~ Major railroads
~ Inland waterways
⛴ Major sea routes
⛴ Major seaports
✈ Major airports

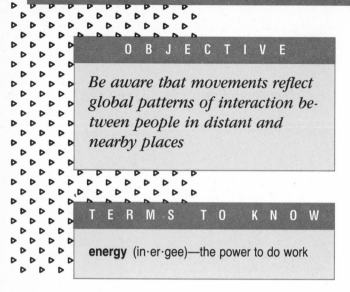

Lesson 2   Major Ocean Trade Routes

OBJECTIVE

Be aware that movements reflect global patterns of interaction between people in distant and nearby places

TERMS TO KNOW

energy (in·er·gee)—the power to do work

Some people feel that the modern world in which we live began on January 10, 1901, on a small hill near Beaumont, Texas. On that day the first big oil strike in history was made. The Spindletop Field, as it was called, poured out more oil than anyone knew what to do with. And soon other oil fields were found. The price of oil fell. Oil became so inexpensive, or cheap, that anyone could afford fuels made from it. Fuels made from oil could be used to run engines in trains, cars, tractors, ships, planes, and factories. The age of cheap **energy** had arrived. Energy is the power to do work. Everything we use is made or brought to us by the use of energy.

How Cheap Energy Changed the World

Cheap energy changed the world. Low-cost energy made it possible to do many kinds of work at a low cost. Farmers using tractors instead of horses to pull their plows could work more land and grow more food. Factories could make goods at a lower cost. Cheap gasoline let ordinary people travel far from home. And soon many new products were being made from oil. These new products made people's lives easier.

Oil is still the single most important fuel. It is perhaps the best example of how people in the world are interdependent. Countries which use more oil than they produce depend on oil from other countries. Countries which have oil to sell depend on the money they earn from oil to buy goods, build roads and schools, and do hundreds of other things.

Oil is very important. This makes it very valuable. But remember that our modern world was built on low-cost energy. In order for everyone to be able to afford the things that make life easier, the price of these things must be kept as low as possible. It costs money to carry oil from the countries that have it to those that need it.

The cheapest way to carry large amounts of goods is by water. Huge ships have been built to carry oil. The paths these ships follow as they cross the oceans can tell us much about the oil trade.

Using Your Skills

A   Use Map 4–3 to answer these questions.

1. What do the shaded lines and arrows on the map show? _____

2. Why are some of the shaded lines and arrows wider than others? _____

_____

3. Where does most of the oil shipped by ocean come from? _____

4. To what three main places does most of the oil shipped by ocean go? _____

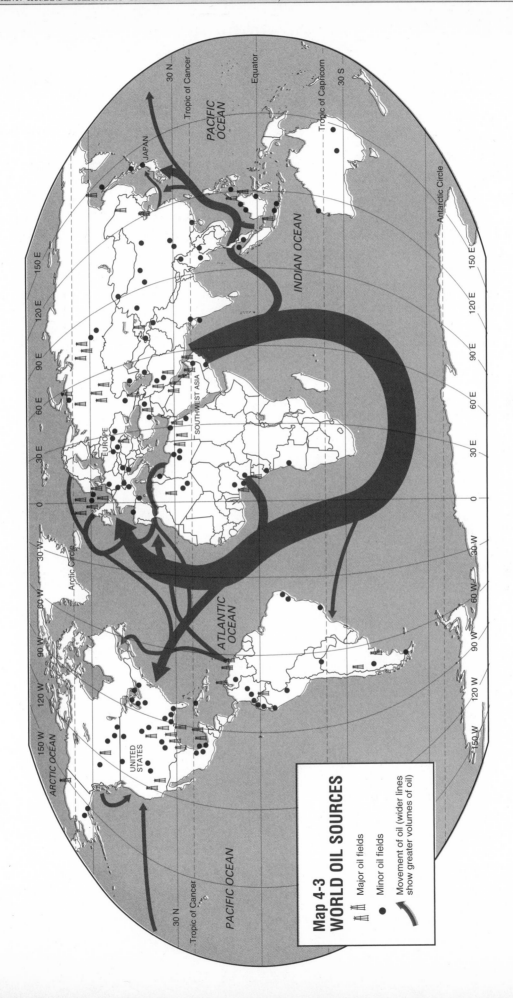

Map 4-3
WORLD OIL SOURCES

- ⚒ Major oil fields
- ● Minor oil fields
- ➤ Movement of oil (wider lines show greater volumes of oil)

B   Use Map 4–4 to answer these questions.

1. Which part of the world does the United States trade with the most? How can you tell?   The United States

trade most with Europe. The width of the line between the United States and Europe is wider than any lines.

2. How does this map show the importance of the Panama Canal?   The lines showing ocean traffic through Panama are

wide.

C   Use both maps in this lesson (4–3 and 4–4) to answer these questions.

1. What differences in the two maps can you see with regard to ocean traffic involving the United States?   All

the lines on the map showing movement of oil go to the United States. The map of world ocean traffic shows many more lines going to

and from the United States.

2. Why are there lines connecting more different countries on the map of world ocean traffic than on Map 4-3?

The map of world ocean traffic shows the movement of all kinds of goods. The other map shows only the movement of oil.

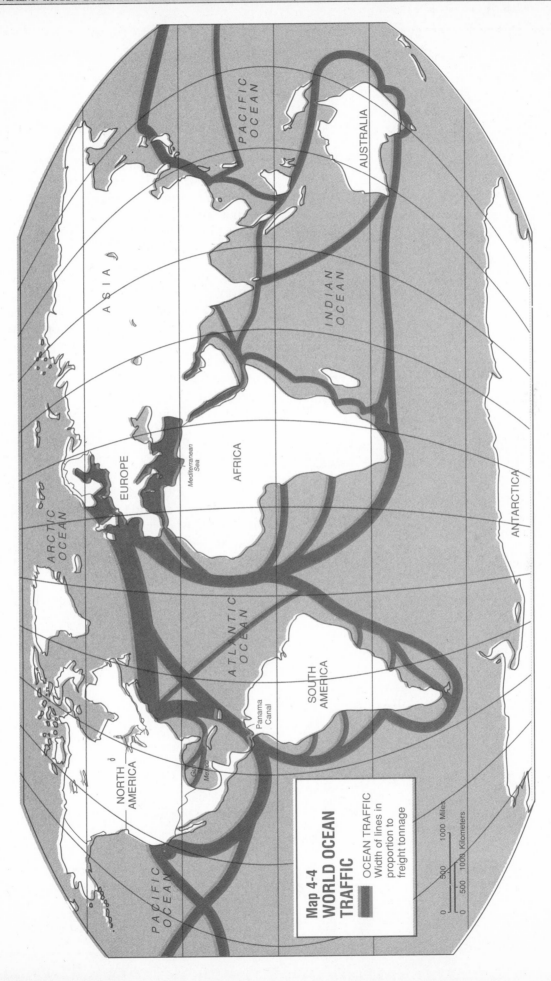

Map 4-4
WORLD OCEAN TRAFFIC

OCEAN TRAFFIC
Width of lines in
proportion to
freight tonnage

1000 Miles

1000 Kilometers

500

500

0

0

PACIFIC OCEAN

ASIA

ARCTIC OCEAN

EUROPE

Mediterranean Sea

AFRICA

INDIAN OCEAN

AUSTRALIA

ANTARCTICA

ATLANTIC OCEAN

SOUTH AMERICA

Panama Canal

NORTH AMERICA

Gulf of Mexico

PACIFIC OCEAN

Lesson 3   Cultural Expansion

OBJECTIVE

Describe ways in which people move their ideas across the earth

TERMS TO KNOW

culture (KUHL·cher)—the way of life of a people

S omeone once said that "There is nothing so powerful as an idea whose time has come." The person meant that when people are ready to accept a new idea, great changes can take place.

The world in which we live has been greatly affected by the spread of ideas. With those ideas have come many other things: religions, foods, types of clothing, laws, and languages, to name a few. The greater part of the earth's surface has been affected by the spread of **cultures.** Culture is the way of life of a group of people. Two cultures largely shaped the world we live in today: the European and the Islamic.

The Islamic Culture

The Islamic culture began in the Middle East. It started with a religion, Islam, which was founded by an Arab trader named Muhammad. Muhammad died in the year 632 A.D. Then his followers, called Muslims, began to spread Islam to other lands. In little more than a hundred years Muslims ruled countries from Spain to India. Later Islam spread still farther, as you can see from the map.

The Muslims spread their religion by force. However, many people were willing to accept Islam. Islam did not require people to give up their whole way of life. Any person could become a Muslim. Many people joined the new religion. They brought their own ideas and way of life with them. Muslims borrowed ideas from the many cultures of the lands they ruled. The Islamic culture became a mixture of the ideas from many different peoples.

Location played an important part in the spread of Islam. Look at Map 4-5 which shows the spread of Islam. The Islamic countries were between Europe, to the east, and India and China, to the west. At this time Europe was made up of many small lands ruled by kings. Most of the people were poor, and life was hard. At this same time, China was the center of a great culture which had books, fine silk, spices, gold, and many other riches.

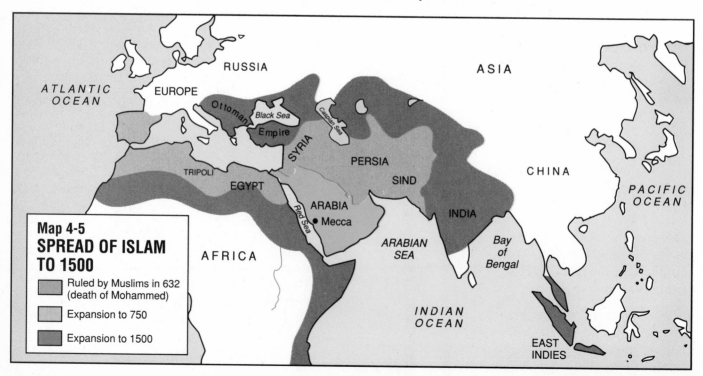

Map 4-5
SPREAD OF ISLAM TO 1500

Ruled by Muslims in 632 (death of Mohammed)

Expansion to 750

Expansion to 1500

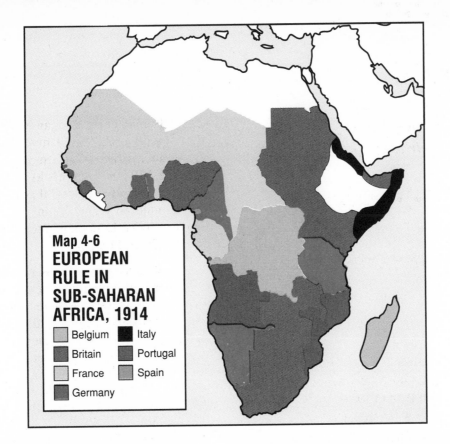

Map 4-6
EUROPEAN RULE IN SUB-SAHARAN AFRICA, 1914

- Belgium
- Britain
- France
- Germany
- Italy
- Portugal
- Spain

A traveler from Europe named Marco Polo visited China from 1271 to 1295. After he returned to Europe, he wrote a book about his trip. People from Europe wanted the riches of China. Trade began. As you can see from the map, this trade had to go through the hands of the Muslims. The Muslims became rich and very powerful from the profits they made from this trade. This wealth and power allowed them to spread their culture.

Much of the world today is still influenced by Muslim people. Southwest Asia, India, Africa North of the Sahara, and Southeast Asia are all areas where the Islamic culture is still strong.

The European Culture

The spread of European cultures came later than the spread of Islamic culture. In fact, the spread of European cultures owes much to the Muslims. It was through the Muslims that Europeans learned about Chinese inventions such as paper, printing, the compass, and gunpowder. These inventions helped Europeans grow more powerful and spread their cultures all over the world.

The compass, for example, helped European sailors find new trade routes to China. Once traders from Europe could sail directly to China from Europe, they no longer had to buy goods from the Muslims. All the

profits went to Europeans. This helped European countries become stronger. Paper and printing meant that Europeans could write down their ideas and send them anywhere—in books. And having many copies of books, printed on a printing press, instead of just a few copies written by hand, meant that ideas could spread much faster.

Gunpowder gave the Europeans a great advantage over people who were still using bows and arrows. With guns and cannons the Europeans took land after land. More than that, they took whole continents: North America, South America, Africa, Australia.

The most important way Europeans spread their cultures to new lands was by settling there themselves. After 1800, Europeans moved to other lands by the millions. Between 1900 and 1914, almost a million people left Europe each year. This was the greatest movement of people in all of history.

With the Europeans came their languages, their religions, and their customs. The culture of the United States, for example, is largely borrowed from that of England. In turn, the American culture is similar to the Canadian and the Australian, which also came from the English. The culture of much of Latin America comes from Spain and Portugal. As Map 4–6 shows, many African countries owe part of their culture to European countries which once ruled them.

Using Your Skills

A   Answer these questions.

1. What is culture?   a group of people's way of life

   _____

   _____

2. Describe the areas covered by the Islamic culture by 1500.   North Africa, Spain, part of southern Europe and Asia,

   part of India, and the East Indies

   _____

   _____

3. What part did location play in the spread of the Islamic culture?   The Muslims were located between Europe and

   China. The money they made from the trade between Europe and China helped them spread their culture.

   _____

   _____

4. How did the spread of the Islamic culture help the spread of European cultures?   Europeans learned of Chinese

   inventions such as gunpowder, paper, printing, and the compass through Muslims. The Europeans used these inventions to spread their

   culture.

   _____

   _____

5. What was the most important way the Europeans spread their cultures?   _____

   The Europeans moved to new lands to live. They took their culture with them.

   _____

   _____

6. What were the only two parts of Sub-Saharan Africa *not* ruled by Europeans in 1914?   Liberia and Ethiopia

   _____

   _____

Lesson 4 The Triangular Trade

OBJECTIVE

Know that movements are concentrated in some areas and sparse in other areas, thereby creating patterns of centers, pathways, and hinterlands

TERMS TO KNOW

hinterland (HINT-er-land)—area where raw materials are grown or gathered

manufactured good (man-yoo-FAK-cherd gud)—thing made from raw materials

profit (PRAH-fit) the money left over after all expenses are paid

raw material (raw muh-tir-ee-uhl)—thing from which other things can be made

triangular trade (TRY-ang-yoo-ler trayd)— trade between New England, England, the West Indies, and Africa in the 1700s

Imagine that you are flying high above the earth. As you look down, you see towns and cities scattered across the countryside. Each town or city has a downtown area with buildings close together. Around it are homes and other buildings that are farther apart. Outside the city, buildings get farther and farther apart. In some places it may be miles between buildings. Roads and perhaps railroads connect the towns with each other and with the buildings in the countryside. As you look down, you see bright, moving flashes of light. It is the sun reflecting off cars and trucks moving on the roads. Most of the cars and trucks are moving to and from the cities.

Centers and Pathways of Trade

What you see from high in the air is a model of how people trade with each other. The cities are centers of trade. These centers are where goods are bought and sold. The areas surrounding the cities are where the goods are produced. The roads and railroads are the pathways over which the goods are carried to the cities to be sold.

If you could fly higher still, so that you could see a large part of the earth's surface at once, you would see much the same thing. It would be easy to pick out the countries where most trade was taking place. A great many ships would be moving to and from some countries. Few ships would be moving to and from other countries. Two main things make up most trade. The first is **raw materials.** Raw materials are things like oil, wood, coal, iron ore, and cotton. Useful things can be made from raw materials. For example, cotton can be made into thread and cloth. These can then be used to make clothing. The things made from raw materials are called **manufactured goods.** Some examples of manufactured goods are cars, radios, shoes, and potato chips.

Some countries produce mostly raw materials. Others produce mainly manufactured goods. As you might expect, there is a great deal of trade between such countries. The countries which have raw materials sell them to make money with which to buy manufactured goods. The countries which make goods must buy raw materials to make those goods. The manufactured goods are then sold to the countries which need them.

Each time raw materials or manufactured goods are sold, the seller makes a **profit.** Profit is the money left over after all expenses are paid. Trade takes place because people can make a profit by buying and selling goods.

The Triangular Trade

The early history of the United States provides an example of how trade works. In the 1700s the land we know as the United States was owned by England. The land was rich in raw materials such as fish, lumber, and cotton. England also owned many of the islands southeast of the United States called the West Indies. Large farms on these islands grew sugarcane. England had many factories. It made the countries it owned buy all their manufactured goods from England.

Trade between England and the lands it owned worked this way. Ships from the northeastern part of what is now the United States, called New England, would carry fish to the West Indies. There the fish was sold to feed the enslaved Africans who worked on the sugar farms. Sugar farmers paid for the fish with sugar. Ships then took the sugar to England. There it was sold or traded for manufactured goods. The manufactured goods were then taken back to New England. The goods were then sold and the money used to buy more fish.

This trade was called the **triangular trade** because the ships moved in a triangle from New England, to the West Indies, to England, and back to New England.

The triangular trade worked in another way, too. It would begin in New England. Fish would be taken to the West Indies and traded for sugar. Then the sugar would be taken back to New England. There it would be made into rum. Rum is a drink with alcohol in it. The rum was taken to Africa's Gold Coast. There it was traded for enslaved Africans. The enslaved were taken to the West Indies and sold to the sugar farmers.

As you can see, the triangular trade had centers and pathways. The centers were the cities in New England, the West Indies, and England where goods were bought and sold. The pathways were the routes the ships followed across the ocean. The triangular trade also had areas where raw materials were grown or gathered. These areas are called **hinterlands.**

Using Your Skills

A   Match each term in Column A with its meaning in Column B.

Column A	Column B
_____ **1.** raw materials	a. the money left over after expenses are paid
_____ **2.** hinterlands	b. trade between New England, England or Africa, and the West Indies
_____ **3.** profit	c. things made from raw materials
_____ **4.** manufactured goods	d. areas where raw materials are produced
_____ **5.** triangular trade	e. things from which useful products can be made

B   Use information from the lesson to complete Map 4–7 of the Triangular Trade

1. Draw the pathways ships followed in the two kinds of triangular trade.

2. Label each pathway with the product ships carried between centers.

3. Label the hinterlands with the product which came from each.

Map 4-7 TRIANGULAR TRADE

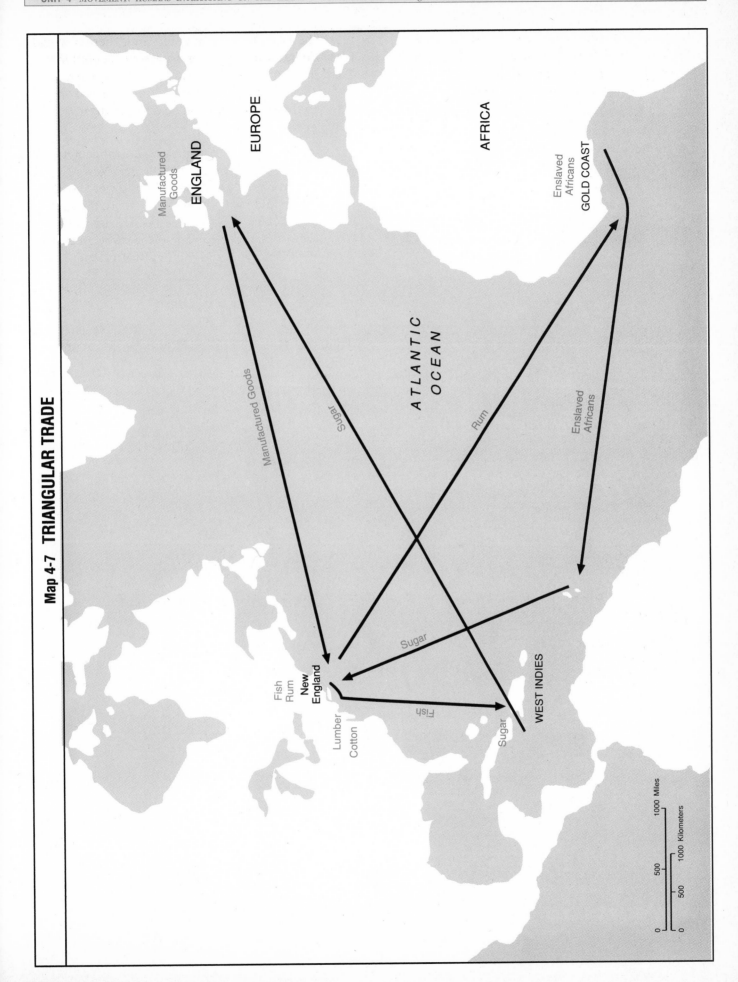

EUROPE

ENGLAND

Manufactured Goods

AFRICA

Enslaved Africans
GOLD COAST

ATLANTIC OCEAN

Manufactured Goods

Sugar

Rum

Enslaved Africans

Sugar

Fish
Rum
New England

Lumber
Cotton

Fish

Sugar

WEST INDIES

1000 Miles

1000 Kilometers

500

500

0

0

OBJECTIVE

Know that few places are self-sufficient and most rely on trade

TERMS TO KNOW

balance of trade (BAL·uhns uhv trayd)—the difference in value between a country's imports and exports

export (EK·sport)—thing a country sells

import (IM·port)—thing a country buys

self-sufficient (self suh·FISH·uhnt)—able to meet all of one's own needs

Look around you. Notice all the different things you see—walls, furniture, the clothing people are wearing. Outside are cars and trucks, and perhaps someone using a lawn mower. Have you ever wondered where all these things come from? Where was your desk made? And where did the raw materials to make the desk come from? Could you walk outside and find all the wood, metal, and plastic needed to make a desk? What about the other things all around you?

Almost anything that you buy will have a label or mark that tells where it was made. Your ballpoint pen may have been made in Japan, your shoes in Italy, your shirt in Taiwan, and your belt in Mexico. Some things, of course, will have been made in the United States. But chances are you can easily find things you own that were made in a dozen different countries.

Why the United States Trades with Other Countries

The United States is one of the richest countries on earth. It has many natural resources. It has many factories that make huge amounts of goods. But the United States is not **self-sufficient.** To be self-sufficient means to meet all of one's own needs. The United States cannot meet all its own needs. For example, even though the United States produces a great deal of oil, it uses more than it can produce. The

rest is bought from other countries. The United States must also buy things it cannot produce. For example, the climate in the United States will not allow such things as coffee, tea, bananas, and rubber to be grown. These things must be bought from countries that can grow them.

The things that a country buys are called **imports.** The things that a country sells are called **exports.** No other country buys and sells more things than the United States. The most important United States imports are oil, machinery, cars, metals, and foods that cannot be grown in the United States. The most important exports are machinery such as computers, chemicals, metal goods, and farm products such as wheat.

The Balance of Trade

How do you suppose the United States pays for the things it buys from other countries? If you said by using the money it gets from selling other things, you are right. Suppose you are a wheat farmer in Kansas whose wheat is sold to Russia. You can take the money you got for the wheat and buy a new pickup truck made in Japan.

All the money that a country makes from the sale of exports and the money spent on imports is added up each year. The difference is called the **balance of trade.** Some countries export more than they import. These countries are said to have a favorable balance of trade. In other words, a country with a favorable balance of trade sells enough to other countries to pay for all the things it imports and have money left over.

Other countries buy more than they sell. These countries are said to have an unfavorable balance of trade. These countries do not sell enough to other countries to pay for all the things they import. A country which imports more than it exports has to borrow money to pay for the things it buys. Experts disagree about whether an unfavorable balance of trade harms a country.

The United States does not have a favorable balance of trade. Table 4–1 shows the United States balance of trade for three recent years. A minus sign (–) before a number in the "Balance of Trade" column means that the United States bought more than it sold.

United States Balance of Trade

Year	Value of Imports	Value of Exports	Balance of Trade
1991	$487 billion	$422 billion	–$65 billion
1992	$533 billion	$448 billion	–$85 billion
1993	$580 billion	$465 billion	–$115 billion

Source: *World Almanac 1995*

Table 4–1

Using Your Skills

A **Answer these questions about the table "United States Balance of Trade."**

1. What was the value of imports the United States bought in 1991? $487 billion

2. What was the value of exports the United States sold in 1991? $422 billion

3. Was the United States balance of trade favorable or unfavorable in 1991? By how much?   It was unfavorable by $65 billion.

4. How much more did the United States buy than it sold in 1993? $115 billion

B   **Use Graphs 4–1 and 4–2 to answer these questions.**

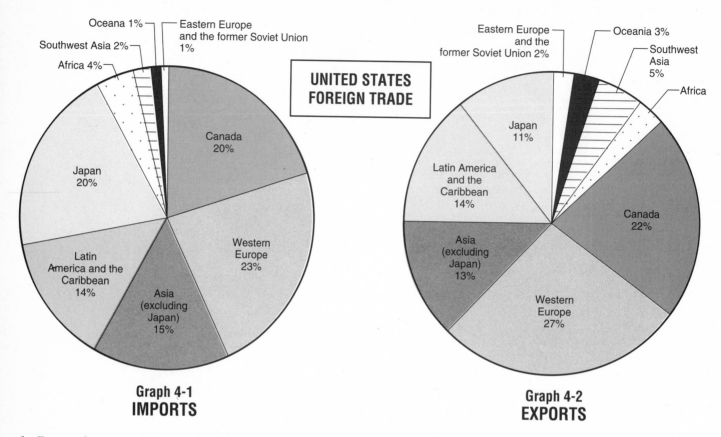

Graph 4-1
IMPORTS

Graph 4-2
EXPORTS

1. From what part of the world does the United States import the largest amount of goods?   <u>Western Europe</u>

2. Which two countries sell the most goods to the United States?   <u>Japan and Canada</u>

3. What part of the world buys the most United States goods?   <u>Western Europe</u>

4. What percentage of United States exports does Japan buy?   <u>11 percent</u>

5. What percentage of United States imports comes from Japan?   <u>20 percent</u>

6. What part of the world buys the smallest percentage of United States exports?   <u>Eastern Europe and the former</u>

<u>Soviet Union</u>

C   **Use the information in the paragraph below to complete Map 4–8. Label each country named. Use shading to show whether the United States balance of trade was favorable or unfavorable. Make a legend to explain the shading.**

The main countries with which the United States had a favorable balance of trade in 1993 were the Netherlands, Eastern Europe and the former Soviet Republics, the United Kingdom, Belgium, Australia, Russia, and Egypt. The United States had an unfavorable balance of trade with Canada, France, Germany, Italy, Saudi Arabia, Japan, China, Taiwan, Korea, Nigeria, Thailand, and South Africa.

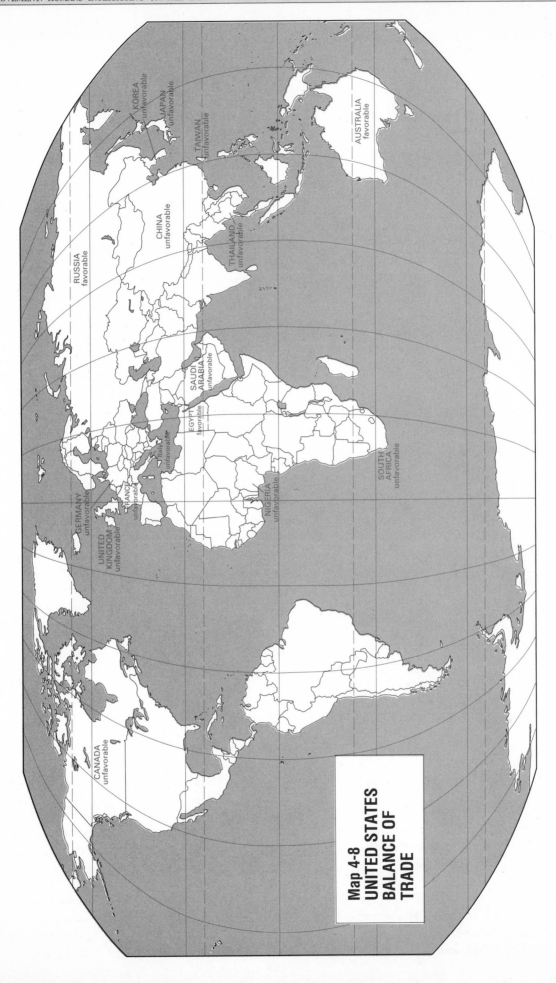

Map 4-8
**UNITED STATES
BALANCE OF
TRADE**

Recognize the relationships between human activities and various locations

I s there a park near your home where you like to go? Where do you go to meet your friends? How far from your home is the store where you shop for food? Do you usually travel by car, on foot, or by train?

It seems that people are always on the move. Most people live in one place, work in another, and shop in a third. In all this travel, Americans depend a great deal on their cars. Think about how you travel. How much time each day do you spend in a car? When was the last time you took a taxi, or a city bus, or a subway? It should not surprise you that the "average" American family takes 97 percent of its trips by car. Only 3 percent of trips are taken by train, streetcar, elevated rail, airplane, taxi, bus, school bus, bicycle, walking, or other form of transportation.

Knowing Your Travel Habits

Of course, no one is "average." Your travel habits probably are quite different from these averages. However, knowing what your travel habits are can help you make decisions about ways to improve them. For example, do you often make several trips from home to various stores in the same day? If so, you may find that you could save time by planning your travel so that you visit all the stores on the same trip. Combining all the trips into one may also save fuel. Thus, planning may help conserve a valuable natural resource.

Effects of Cars on American Life

The American "love affair" with the car has many effects on the way we live. A great many people have jobs building and taking care of all the cars we need. Our nation has spent huge amounts of money to build roads. Air pollution from cars makes the air in many of our cities poor.

The use of cars also means that we have many choices. We can choose to live near work or school, or many miles away. We can travel to distant cities with ease. A trip that would have taken a person on horseback three days a hundred years ago can now be made in two hours. Therefore, we can choose to travel just for fun.

Using Your Skills

A Gather information about your family's travel habits. Find out how far your family travels for each kind of trip listed in Table 4–2. Compare your family's travel habits to the figures for the average family.

Household Travel

Purpose of Trip	Average American Family Average Length in Miles	Your Family Average Length in Miles
Trip to work	8.5	Answers will vary.
Work-related business	11.4	
Shopping	5.2	
Family business	6.7	
School or church	5.5	
Visit friends or relatives	10.8	
Other recreation	8.7	
Vacation	114.0	
Other	7.2	

Table 4–2

B Use your own paper to gather information about and make a map of transportation in your community. Use what you have learned about direction, distance, scale, and map symbols and legends in this book.

1. Find a road map or city street map of your local area. Using the map legend, identify the main highways, roads, and streets shown on the map. Draw those roads and streets on your map.

2. Use interviews, the telephone directory, or information from the Chamber of Commerce to find out what businesses are the largest employers in your community. Plot the locations of these businesses on your map.

3. Use maps or the telephone directory to find the locations of schools, churches, stores, and shopping centers, parks, movie theaters, lakes, rivers, and other recreation spots. Plot these on your map.

4. If your town or city has a bus or passenger train system, draw the routes on your map.   Answers will vary.

C On your own paper, answer the following questions about your map.

1. Does your community have major highways and roads running in all directions, or do most of the main transportation routes run just north-south or east-west?

2. Are major employers, schools, churches, shopping centers, parks, and other important places served by main roads?

3. If your community has a bus or passenger train system, how well do the routes serve the main places people need to go?

Answers will vary.

Know that all cultures are inter-dependent

How would you like to have thousands of people all over the world working for you? You do—right this minute. In fact, even while you sleep, people around the world are working in mines and factories and on farms to produce things you need. As you read this, ships and planes are crossing the oceans to bring these things to you.

You may be feeling rather special by now. But you needn't be. The fact is that all people, all over the world, work for each other. We all depend on others for things we cannot make for ourselves. We are all interdependent.

The Global Pencil

One example of how interdependent we all are may be in your hand right now. Or perhaps it is in your pocket, or your desk, or your notebook.

It takes the efforts of thousands of people from as many as twenty different countries and states to make one little pencil. You can imagine how many people it must take to make something like a car or a television set!

The logs are loaded onto a <u>truck</u>. The truck may have been made in <u>Michigan</u>. However, it could have been put together in a plant just across the border from the United States in <u>Canada</u>. And, of course, the trucks run on fuel made from crude oil.

The logs may be taken to a <u>sawmill</u> in <u>California</u>. The logs are sawed into small pieces before being sent to the <u>factory</u> in <u>Pennsylvania</u>. There the other parts that make up a pencil are added.

The "lead" in pencils is not really lead at all. Pencil lead is a mixture of several things. <u>Graphite</u> comes from mines in <u>Sri Lanka</u>. It takes the work of miners and dock workers in Sri Lanka to put the graphite on a <u>ship</u> built in <u>Japan</u>. The <u>ship owner</u> lives in <u>France</u>. The <u>ship company</u> that operates the ship does business from <u>Liberia</u>. The graphite is mixed with <u>clay</u> from <u>Mississippi</u> and <u>wax</u> from <u>Mexico</u>.

For many people the most useful part of a pencil is the eraser. The <u>rubber</u> in the eraser likely came from <u>Malaysia</u>. The gritty stuff in the eraser that wears the pencil marks off the paper is <u>pumice</u>. Pumice comes from volcanoes in <u>Italy</u>. The piece of metal that holds the eraser in place is made of brass. Brass is made of zinc and copper. <u>Zinc</u> comes mainly from the <u>United States</u>, <u>Canada</u>, <u>Australia</u>, the <u>Soviet Union</u>, and <u>Ireland</u>. The <u>copper</u> may have come from <u>Bolivia</u>, <u>Chile</u>, or <u>Zambia</u>.

The pencil is almost finished. But first it must be painted. One of the main things that goes into the paint is <u>castor oil</u>. Farmers in <u>Africa</u> grow the castor bean plants. After the pencil is painted, the name of the maker is stamped on it. The black paint has <u>carbon black</u> from the far north of <u>Texas</u> in it.

Now the pencil is finished, but it must still be sent to you. Hundreds of other people are involved in shipping and selling the pencil after it leaves the factory. People in any one of the fifty states could have played a part in bringing you the pencil you use every day.

Using Your Skills

A **Read the description below of how a pencil is made. The underlined words tell you what goes into a pencil, or where work on pencils is done. Write each word in the correct location on Map 4–9 on the next page.**

The roar of the chain saw stops as the <u>cedar tree</u> starts to fall. Much of the wood for pencils comes from trees in <u>Oregon</u>. The <u>chain saw</u> may have been made in <u>Japan</u>. The gasoline to run it started out as <u>crude oil</u>.

Perhaps the oil came from <u>Texas</u>. But chances are good that the oil came from several places, such as <u>Mexico</u>, <u>Alaska</u>, <u>Saudi Arabia</u>, or the <u>North Sea</u> off the coast of the United Kingdom.

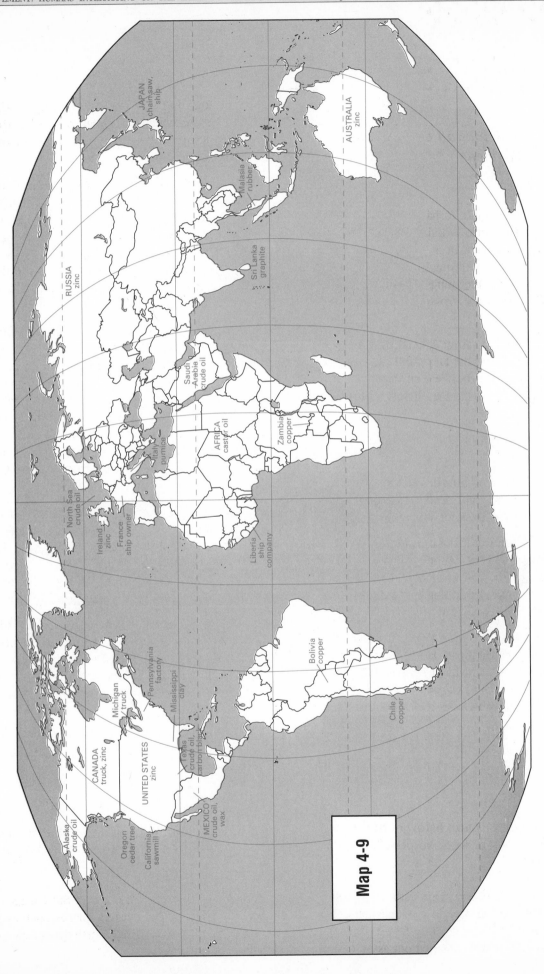

Map 4-9

U N I T   4   R E V I E W

A   Write the correct term from the box in each sentence.

balance of trade	culture	imports
interdependence	exports	trade
transportation	raw materials	profit

1. Moving people and goods from place to place is called _____transportation_____.

2. _____Trade_____ is the buying and selling of goods.

3. People depending on each other is called _____interdependence_____.

4. The way of life of a people is their _____culture_____.

5. Useful things can be made from _____raw materials_____.

6. People try to make a _____profit_____ by buying and selling goods.

7. A country buys the _____imports_____ it needs from other countries.

8. The things a country sells to other countries are called _____exports_____.

9. Countries which export goods worth more than the goods they import are said to have a favorable

_____balance of trade_____.

B   Write the letter of the best answer in the blank.

____c____ 1. Which country has the best transportation system in the world?
  a. Canada   b. Japan   c. United States

____b____ 2. Why is transportation so important to the United States?
  a. The United States has   b. The United States depends   c. People like to travel to new
     over 4 million miles of   on trade.   places on vacation.
     highways.

____a____ 3. Why is cheap energy important to our way of life?
  a. Everything we use is   b. Cheap gasoline lets people   c. Cheap energy keeps the cost of
     made or brought to us   travel far from home.   goods down.
     by the use of energy.

____c____ 4. Much of the oil used in the United States comes from
  a. oil wells in the United   b. countries in Southwest   c. both a and b.
     States.   Asia

_c_   **5.** Two cultures which largely shaped the world we live in today are the
     a. European and South American.
     b. Islamic and American.
     c. European and Islamic.

_b_   **6.** One reason the Muslims were able to spread their culture was because they
     a. did not accept other people's ideas.
     b. traded with people in many parts of the world.
     c. invented paper and printing.

_c_   **7.** The most important way Europeans spread their culture was by
     a. printing many copies of books.
     b. using guns to take land from other people.
     c. moving to new lands to live.

_a_   **8.** The culture of the United States was borrowed mainly from
     a. England.
     b. China.
     c. the Muslims.

_c_   **9.** The centers of trade where goods are bought and sold are the
     a. hinterlands.
     b. countries which have many raw materials.
     c. cities.

_a_   **10.** One part of the triangular trade in the United States during the 1700s was
     a. the trading of rum made in New England for slaves in Africa.
     b. the trading of sugar from New England for fish caught in the West Indies.
     c. the trading of manufactured goods from the West Indies for slaves in New England.

_c_   **11.** Which statement below best describes the United States?
     a. The United States has many natural resources and is one of the richest countries on earth.
     b. The United States is not self-sufficient and must import many of the things it needs.
     c. both a and b.

_a_   **12.** A country pays for the things it needs from other countries by
     a. selling things it has that other countries need.
     b. turning out more things in its factories than its own people need.
     c. saving its raw materials for its own use.

_c_   **13.** The balance of trade in the United States between 1991 and 1993
     a. stayed about the same.
     b. grew more favorable.
     c. grew more unfavorable.

_b_   **14.** The United States gets most of its imports from, and sends most of its exports to,
     a. Japan.
     b. Western Europe.
     c. Canada.

_c_   **15.** The average American family takes what percentage of its trips by car?
     a. 3 percent
     b. 50 percent
     c. 97 percent

_b_   **16.** The making of a pencil is a good example of
     a. self-sufficiency.
     b. interdependence.
     c. triangular trade.

C **Answer these questions about Map 4–10.**

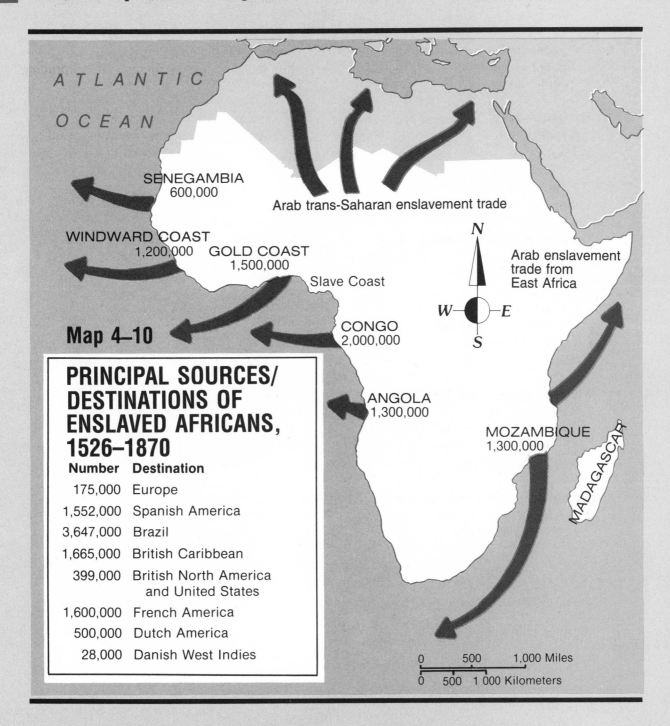

A T L A N T I C

O C E A N

SENEGAMBIA
600,000

Arab trans-Saharan enslavement trade

WINDWARD COAST
1,200,000   GOLD COAST
1,500,000

Slave Coast

N

Arab enslavement
trade from
East Africa

W   E

Map 4–10

CONGO
2,000,000

S

**PRINCIPAL SOURCES/
DESTINATIONS OF
ENSLAVED AFRICANS,
1526–1870**

ANGOLA
1,300,000

MOZAMBIQUE
1,300,000

MADAGASCAR

Number	Destination
175,000	Europe
1,552,000	Spanish America
3,647,000	Brazil
1,665,000	British Caribbean
399,000	British North America and United States
1,600,000	French America
500,000	Dutch America
28,000	Danish West Indies

0   500   1,000 Miles

0   500   1 000 Kilometers

_____ **1.** The greatest number of the enslaved taken from Africa came from the region called the
a. Windward Coast.   b. Gold Coast.   c. Congo.

_____ **2.** The greatest number of the enslaved were taken to
a. Europe.   b. Brazil.   c. United States.

_____ **3.** According to information on this map, the trade of enslaved Africans lasted for at least
a. 100 years.   b. 200 years.   c. 300 years.

UNIT 5 Regions: How They Form and Change

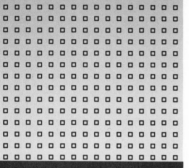

OBJECTIVES

After completing this unit, you will be able to:

- explain how regions may be defined by cultural or physical features, or explain geography of everyday life,

- describe how regions vary in scale,

- name different kinds of regions.

Geographers use regions to help organize the study of geography. Regions are the "tools" of geography. They may be defined any number of ways. Here are the most common types of regions, with examples of each.

A *physical region* is an area that has one main physical feature or environmental characteristic. The Rocky Mountains are a physical region. So are the Sahara, the Alps, and the Gulf-Atlantic Coastal Plain. Climatic regions are physical regions.

Political regions usually have a common political system, such as democracy or socialism. They are surrounded by political boundaries. Brazil is a political region, as are the United States, Poland, and Argentina. In the United States, there are political boundaries around states. Texas is a political region. So are California, Ohio, Florida, and New York.

Uniform regions have a common feature that sets them apart. The Cotton Belt of the United States is a uniform region in which cotton is the main crop. Many cities have areas in which most of the people are of one nationality. These areas are uniform regions.

An example of a *functional region* would be a large city and the surrounding area. People live in the surrounding area and work and shop in the city. Other functional regions include school districts, and supermarkets or shopping centers.

Regions: how they form and change is one of the basic themes of geography. Many aspects of the

Unlucky Thirteen

How many times have you heard that the number thirteen is unlucky? This is a fear that is common all over the world, and one that many people take seriously. Buildings in the United States label the floor that follows twelve as "fourteen" because they have found that people don't want to live or work on the thirteenth floor. The French never issue the house number thirteen, and, in Italy, they leave the number thirteen out of the national lottery.

Not all countries and cultures share the belief that the number thirteen is unlucky. In Korea, for example, you can find a thirteenth floor but not a fourth floor. Since the Korean word for the number four is the same as the Korean word for death, four is considered unlucky in that country.

Amazing Statistics

- In this city, more than 190 people pick up the telephone receiver every second, on average.
- In this city, the people use 2,659,532 quarts of milk and an estimated 7,000,000 eggs a day.
- In this city, a baby is born every 4 minutes and 6 seconds.

These are some pretty amazing statistics or facts. It is obvious from these statistics that the city described must be a huge metropolis filled with people. In fact it is; it's New York City. What is even more amazing is that these statistics applied to New York in 1929. Looking at how it was then, can you imagine what these figures would be like today?

physical and human environment can be used to define regions. Regions may be set up based on language, religion, or educational level of the people. Regions may be based on types of plants, amount of rainfall, or types of soil.

You will recall that geographers use regions to *organize* the study of geography, not make it more confusing. As you can see, it would be very confusing to divide the world into too many regions for study. Therefore, geographers usually divide the world into eight to ten regions based on features such as land, climate, and culture.

In this unit you will not study each of the regions into which the world can be divided. Instead, you will come to understand *how* regions are formed, and how regions can change.

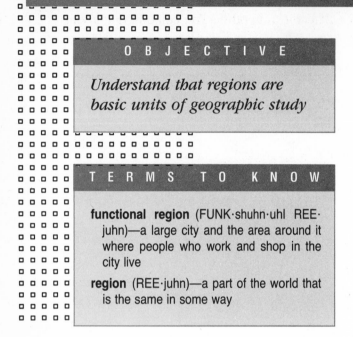

Lesson 1 Communities

Has someone ever said to you, "Can you tell me all about yourself?" A question like this leaves most people with a helpless feeling. Where do you begin an answer to such a question? What should you talk about? What should you <u>not</u> talk about?

Now imagine that someone says to a geographer, "Tell me all about the world." You can see how hard it would be to do this. There are so many different parts of the world to tell about. There are so many different peoples to tell about. A geographer could talk for a very long time and still have more to tell.

Your Community—A Region

Because the world has so many different things to tell about, geographers do not try to study the whole world at once. Instead, they divide the world into **regions.** A region is a part of the world that is the same in some way. Dividing the world into regions makes the study of geography easier. Geographers sometimes choose to make the study of just one region of the world their main work. That way they can learn much about one region, rather than a little about many regions.

There is one region of the world that you probably already know a great deal about. That region is the community where you live. Your community may be a kind of region called a **functional region.** A functional region includes a large city and all the area around it where people live who work and shop in the city. These regions are often called Metropolitan Statistical Areas (MSA's).

Using Your Skills

A Use the directions and questions below and the map form on the next page to make a map of a region—your community.

1. Choose the boundaries of your region. If you live in a Metropolitan Statistical Area, you may use the boundaries of the MSA. If you base the boundaries of your region on political boundaries, you could choose to include the county or town in which you live. Or, you could include only part of your county or town, such as a ward, precinct, or district used to elect government leaders. If you base the boundaries on physical features of the land, you might include only the low or high areas of your community, or the land within a certain distance of a river or highway. If you choose to base your boundaries on cultural features, you might draw your boundary based on the nationality of people who live in an area. For example, your region might include the part of town where most of the people are Italians. Draw the boundary on the map form on the next page.

2. Show the locations of the following features on your map: your home, your best friend's home, the stores where your family shops, your school, and parks.

3. Use shading to show on your map any large areas of land that are used for the same purpose, such as farming, airports, factories, apartments, or stores. Make a legend to show the meaning of the shading.

4. Use the grid on the map to make an index for the map.

5. On your own paper describe the region you have mapped. What are its boundaries? Why did you choose those boundaries? What are the features that make this region different from others you could have chosen?

Answers will vary.

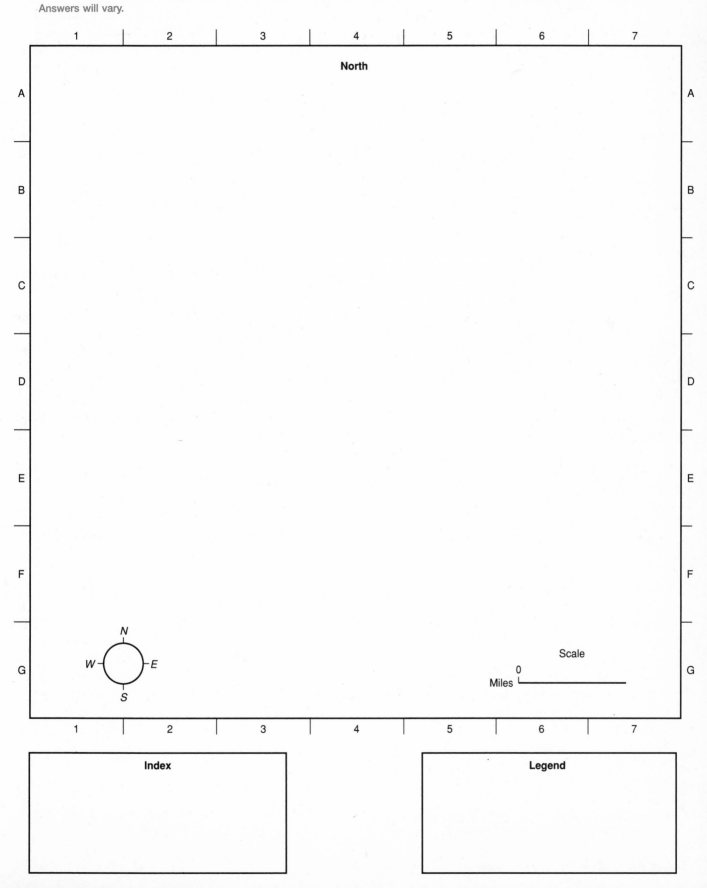

Lesson 2   Physical Regions

Have you seen a picture of earth taken from far out in space? From a great distance, only broad features of the earth stand out. You can see oceans and the shape of the largest landmasses, called **continents.**

Of course, the earth has more features than just oceans and continents. The features of the earth's surface are called landforms. You learned about landforms in Unit 2 of this book. The surface of the earth can be divided into regions based on landforms. Three of the main landforms are mountains, plateaus, and plains. Map 5–1 shows some of the main mountain ranges, plateaus, and plains areas of the earth.

156

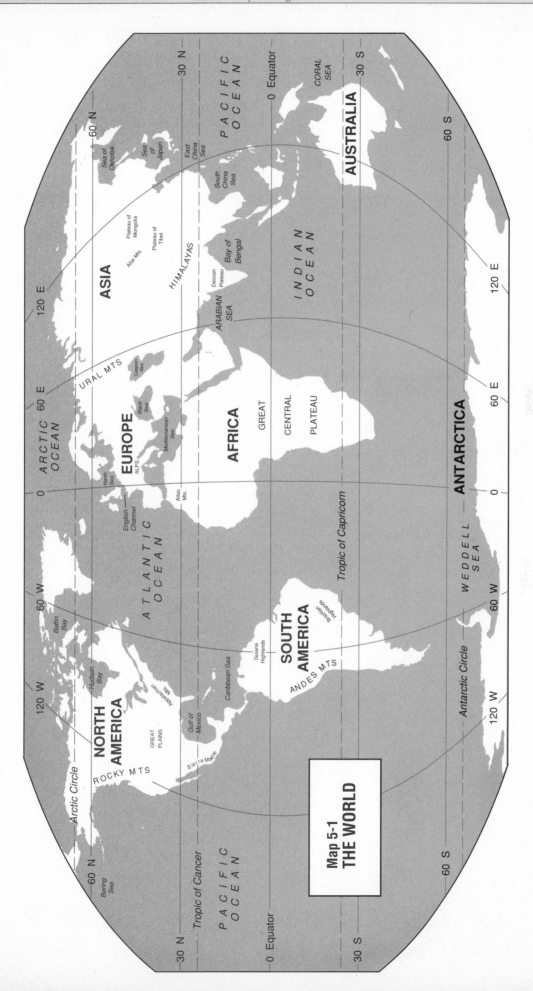

Map 5-1
THE WORLD

Physical Regions

You have read that a region is a part of the earth that is alike in some way. Think of where you live. Is most of the land flat and not too far above sea level? If so, you may live in a plains area. If the land is flat and high above sea level, you may live on a plateau. And of course, if the land around you is very high with many peaks and valleys, you may live in the mountains.

Physical features such as plains, plateaus, and mountains can be used to divide the earth's surface into regions. Map 5–2 shows the physical regions of North America. The boundaries of countries are shown only to help you understand where the regions are in relation to where you live.

Using Your Skills

A   **Use what you have learned in this book about reading maps to answer these questions about Maps 5–1 and 5–2.**

1. What continent is located at latitude 30°S, longitude 120°E?   _____Australia_____

2. What continent is located directly north of the continent of Africa?   _____Europe_____

3. North America is not directly north of South America. What term best describes the direction you would

   travel in going from the Brazilian Highlands to the Great Plains?   _____northwest_____

4. Which physical region of North America runs the greatest distance from north to south?   _the Western Interior_

   Mountains and Basins

5. Which physical region of North America is located immediately to the east of the Western Interior Mountains

   and Basins in the United States?   _the Continental Interior Plain_____

6. Based on the map of physical regions of North America, where would you expect to find more low, flat land:

   along the east coast of the United States, or along the west coast? Why?   _You would expect to find more low, flat_

   land along the east coast of the United States. The map shows that this area is part of the region called the Coastal Plain. The west

   coast is part of the region called the Pacific Mountains and Valleys.

   _____

   _____

   _____

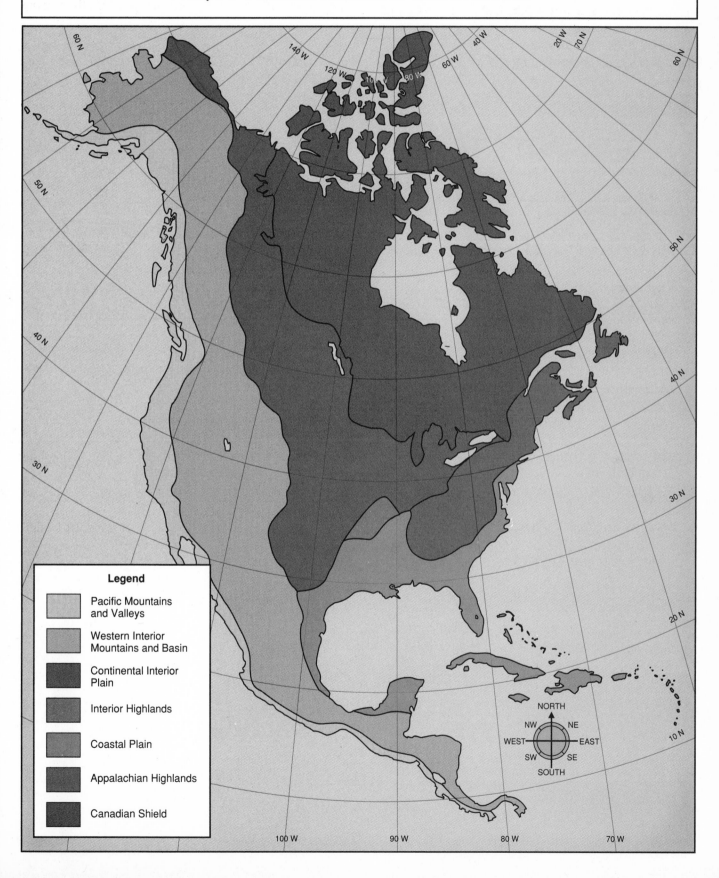

Map 5-2 PHYSICAL REGIONS OF NORTH AMERICA

Legend

- Pacific Mountains and Valleys
- Western Interior Mountains and Basin
- Continental Interior Plain
- Interior Highlands
- Coastal Plain
- Appalachian Highlands
- Canadian Shield

TERMS TO KNOW

democracy (duh·MAHK·ruh·see)—a government in which laws are made by leaders elected by all the people

political region (puh·LIT·uh·kuhl REE·juhn)—an area that has a particular kind of government

D o you know what kind of government the United States has? In the United States, laws are made by leaders elected by all the people. We call this kind of government a **democracy.** Some other democracies are Canada, the United Kingdom, India, and Australia.

Political Regions

One way to define a region is by the kind of government a country has. For example, the United States is a **political region.** A political region is an area that has a particular kind of government. Each country in the world is a political region. The boundary around each political region is called a political boundary. Each state in the United States has a political boundary around it. Each state is also a political region. Each county within each state is a political region. And each voting district within each county is a still smaller political region.

As you can see, the same area can be part of several different political regions. A voting district can be part of a county, a state, and a country all at the same time. In the same way, a country can be a political region all by itself, or it may be part of a region which includes other countries.

Using Your Skills

A The Arab League was formed to help countries ruled by Arabs work together. The countries of the Arab League form a political region. This powerful region is often at odds with the United States. A list of Arab League countries follows. Write the name of each country in the proper place on Map 5–3.

Map 5-3 THE ARAB LEAGUE

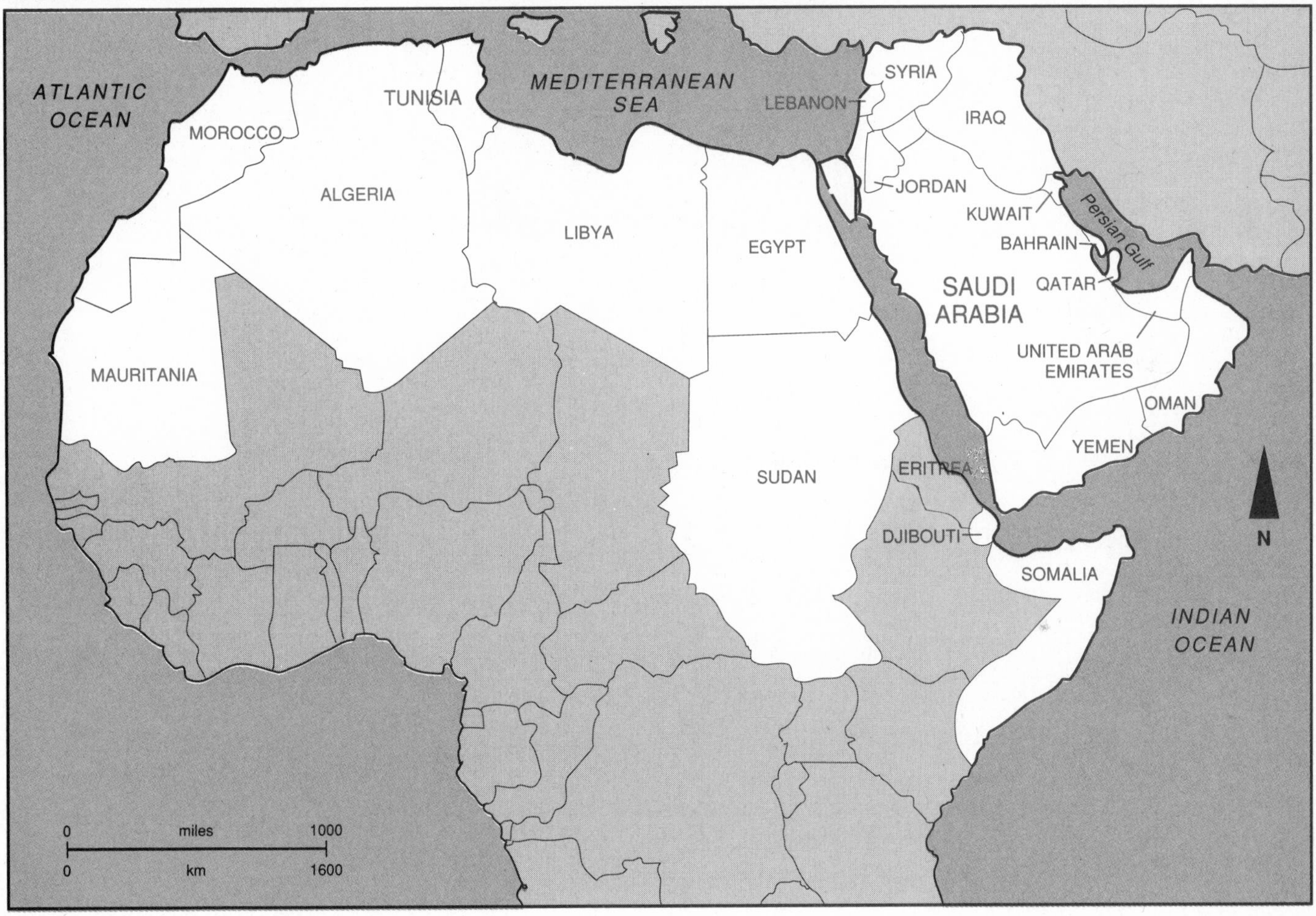

ATLANTIC
OCEAN

MEDITERRANEAN
SEA

TUNISIA

MOROCCO

ALGERIA

LIBYA

EGYPT

MAURITANIA

SUDAN

SYRIA

LEBANON

IRAQ

JORDAN

KUWAIT

BAHRAIN

Persian Gulf

SAUDI
ARABIA

QATAR

UNITED ARAB
EMIRATES

OMAN

YEMEN

ERITREA

DJIBOUTI

SOMALIA

INDIAN
OCEAN

N

| 0 | miles | 1000 |
| 0 | km | 1600 |

Algeria	Bahrain	Eritrea
Egypt	Iraq	Djibouti
Kuwait	Lebanon	Jordan
Mauritania	Morocco	Libya
Oman	Qatar	Saudi Arabia
Somalia	Yemen	Sudan
Syria	Tunisia	United Arab Emirates

B One of the major forms of government in the world has been communism. Under communism, all the people do not elect the leaders who make the laws. Only a few people decide how the country will be run. The United States has tried to stop the spread of communism. However, many countries in the world today have or have had Communist governments. The United States sometimes has trouble with these countries. A list of former Communist and Communist countries follows. Write the name of each country in the proper place on Map 5–4.

Albania	Bulgaria	Yugoslavia
Poland	Romania	Czechoslovakia
North Korea	Hungary	East Germany
China	Tibet	Vietnam
Cuba	Cambodia	Laos
Afghanistan	Soviet Union	

C   Answer these questions.

1. What is a political region?   an area that has a particular kind of government

2. Could a member of the Arab League also be a member of the Communist World?   None are at the present time.
However, if an Arab League country changed to a Communist form of government, it would also be a member of the Communist World.

3. Name all the political regions of which your city or town is a part.   Answers will vary but should include the nation,
state, county and city (where applicable).

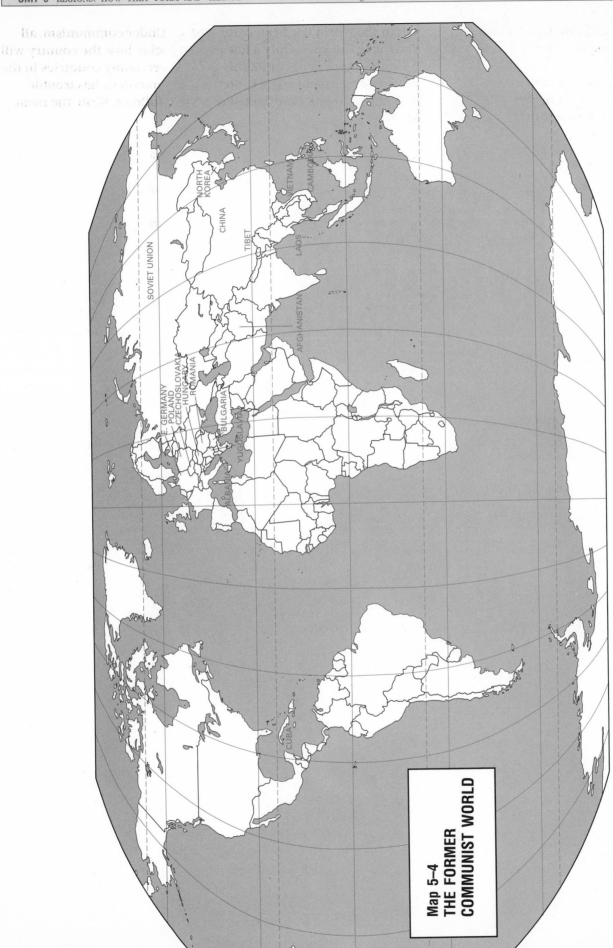

Map 5-4
THE FORMER COMMUNIST WORLD

Lesson 4 Culture Regions

O B J E C T I V E

Explain how regions may be defined by cultural features

T E R M S   T O   K N O W

Communist World (KAHM-yoo-nist world) —region formerly made up of countries with a Communist form of government

culture region (KUHL·cher REE·juhn)—region based on many factors of how people live

developed nation (di·VEL·upt NAY·shuhn) —country which uses a great deal of technology

economic system (ek·uh·NAHM·ik SIS· tuhm)—the way a people produce, get, and use goods and services

free enterprise (free IN·ter·pryz)—an economic system in which people are free to decide what kind of work they will do, and to own businesses and keep the profits

socialism (SOH·shuhl·iz·uhm)—economic system in which the government decides what kind of work people will do, and owns most of the businesses

Third World (therd werld)—region made up of countries that do not take sides with either the Communist World or the Free World

Have you ever traveled to a place where the people were very different from you? What made them different? Was it the language they spoke? The clothes they wore? Did they use money that was strange to you?

A people's way of living is called culture. Culture includes many things. Some of the most important are language, religion, government, the use of technology, and type of economic system.

Elements of Culture

Language is an important part of culture. People who speak the same language feel that they are part of the same group. Think how you feel when you hear people talking in a language you cannot understand. Speaking different languages keeps people apart. Speaking the same language draws people together. Geographers sometimes divide the world into regions based on the languages that people speak.

Another part of culture is religion. There are many different religions in the world. In some countries, most of the people follow one religion. The world can be divided into regions based on the religions that people follow.

The kind of government a people has is an important part of culture. In some countries, people make their own laws. In other countries the rulers make the laws. In lesson 3 of this unit you saw an example of how the world can be divided into regions based on the kind of government people have. The countries that had a Communist form of government can be put in a region formerly called the **Communist World.** Great changes in Communist Europe have taken place in the 1990s. Often you will hear of a region called the **Third World.** The Third World is made up of countries that did not take sides with either the Communist World or the Free World.

You have read that technology is the use of tools and skills to make life easier. In some parts of the world, people use a great deal of technology. The United States, Canada, and Japan are examples of such countries. Some people use very little technology. Countries which use a great deal of technology are sometimes grouped in a region called **developed nations.** Countries which use little technology are called developing nations.

The way a people produce, get, and use goods and services is called their **economic system.** In some countries, people are mostly free to decide for themselves what kind of work they will do. In such countries, a person can own a business and keep the profits from that business. This kind of economic system is called **free enterprise.** The United States is one of the leading free enterprise nations.

In other countries, the government decides what kind of work people will do. In such countries, the

government owns most businesses. This kind of economic system is called **socialism.**

Culture Regions

It is possible to divide the world into regions based on any one of the factors above plus many others. Any one country could fit into many different regions.

However, the study of the world would be very confusing if regions kept changing all the time. One way to make things simpler is to group the countries of the world into broad regions that are based on a combination of many factors. **Culture regions** are based on the factors discussed in this lesson. Using these factors, it is possible to group the countries of the world into just ten culture regions.

Using Your Skills

A   Choose the correct term from the pair in parentheses in each sentence. Underline the correct term.

1. Great changes took place in the 1990s in Europe in the (Communist World, Third World).

2. A country which uses a great deal of technology is called a (developed country, developing country).

3. The way a people produce, get, and use goods and services is called their (culture region, economic system).

4. In a free enterprise system, businesses are owned by the (people, government).

5. The United States is a leader in the (free enterprise, communist) system.

B   Map 5–5 below shows the world divided into ten culture regions. Also below is a list of countries. Write the name of each country on the list in the correct place on the map.

Argentina	Brazil	Canada
Venezuela	Colombia	United States
France	Ireland	Italy
Spain	United Kingdom	Germany
Germany	Poland	Russia
Egypt	Iran	Iraq
Israel	Saudi Arabia	Ethiopia
Kenya	Nigeria	South Africa
Zaire	India	Pakistan
China	Japan	Taiwan
Philippines	Vietnam	Australia

C   Answer the following questions about Map 5–5.

1. Japan is a part of what culture region?   China and East Asia

2. Germany is a part of what culture region?   Europe

3. Russia is a part of what culture region?   Russia and the Independent Republics

4. Mexico is a part of what culture region?   Latin America and the Caribbean

5. Iraq is a part of what culture region?   North Africa and Southwest Asia

6. This map shows three continents each of which are entirely made up of only one culture region. What are the names of these three continents? Of what culture region is each continent a part?   Antarctica: Antarctica, Australia, and Oceania; Australia: Antarctica, Australia, and Oceania; South America: Latin America

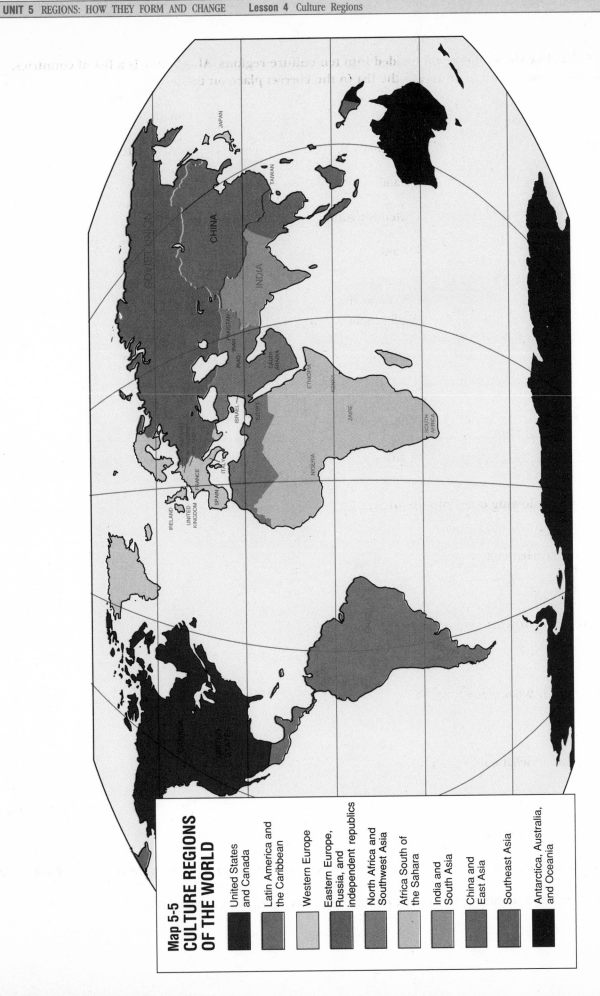

Map 5-5
CULTURE REGIONS OF THE WORLD

United States and Canada

Latin America and the Caribbean

Western Europe

Eastern Europe, Russia, and independent republics

North Africa and Southwest Asia

Africa South of the Sahara

India and South Asia

China and East Asia

Southeast Asia

Antarctica, Australia, and Oceania

Lesson 5   Uniform Regions

OBJECTIVE

Give examples of uniform regions

TERMS TO KNOW

uniform region (YOO·nuh·form REE·juhn) —an area that has one feature that sets it apart

Have you ever walked or driven through a part of a town where most of the people were from Italy, or Greece, or some other country? Perhaps you have been on a trip that took you through mile after mile of desert. Or perhaps you have been to a place where wheat or corn grew in all directions as far as you could see. If you have been to any such place, you know what a **uniform region** is. A uniform region is an area that has one feature that sets it apart.

A uniform region can be set up based on almost any feature. For example, the Cotton Belt is a part of the United States where cotton is the main crop. The Sun Belt is the part of the United States where temperatures are warm most of the year, and snow seldom falls. The *barrios* are parts of cities where most of the people are Hispanic.

Using Your Skills

A   **Answer these questions about Maps 5–6 and 5–7.**

1. What uniform region is shown on Map 5–6?   the Corn Belt

2. What would you expect to see growing on most farms in Iowa, Illinois, and Indiana?   corn

3. What uniform region is shown on Map 5–7?   the Wheat Belt

4. What would you expect to see growing on most farms in Kansas and North Dakota?   wheat

5. About how many miles does the Corn Belt stretch from east to west?   850 miles

6. What is the southernmost state in the Wheat Belt?   Texas

7. In which belt is southern Michigan?   the Corn Belt

8. Which states have land in both the Corn Belt and the Wheat Belt?   Minnesota, South Dakota, Nebraska, Missouri

Map 5-6 THE CORN BELT

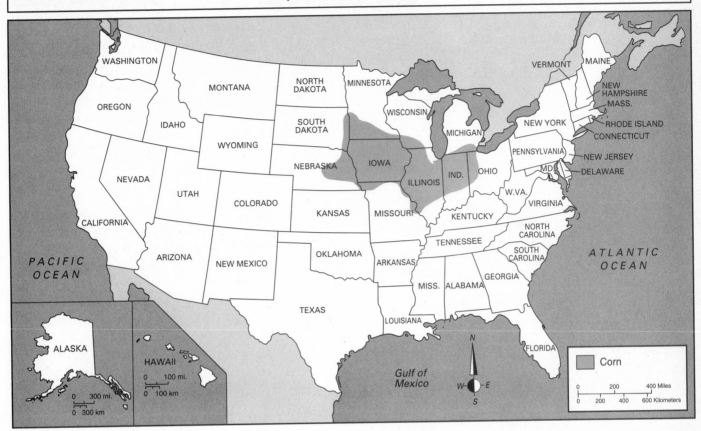

Map 5-7 THE WHEAT BELT

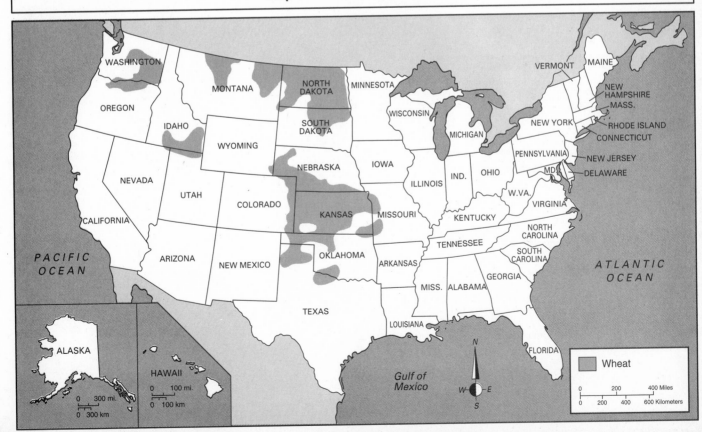

B   Some countries depend on one product for most of the money they earn from exports. Such countries can be grouped into uniform regions according to the product upon which they depend. Below is a list of such countries in Africa. Choose a color or shading to stand for each product. Color or shade each country on Map 5–8. Complete the legend to identify the uniform regions.

Oil		Diamonds	Coffee	Iron Ore
Algeria	Libya	Botswana	Burundi	Liberia
Angola	Nigeria	Lesotho	Rwanda	Mauritania
Congo	Tunisia	Sierra Leone	Uganda	
Gabon				

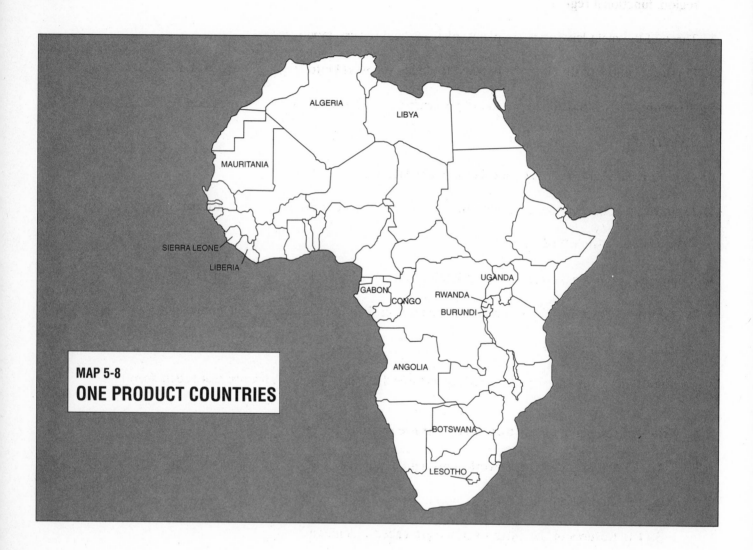

MAP 5-8
ONE PRODUCT COUNTRIES

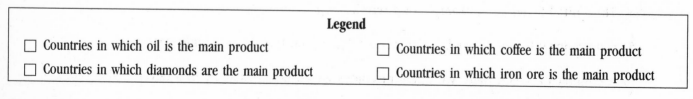

Legend	
☐ Countries in which oil is the main product	☐ Countries in which coffee is the main product
☐ Countries in which diamonds are the main product	☐ Countries in which iron ore is the main product

Answers: Legend should be completed to show color used on map for each region.

U N I T   5   R E V I E W

A   **Underline the term in parentheses in each sentence which will complete the statement correctly.**

1. A region is a part of the world that is (<u>alike</u>, different) in some way.

2. A large city and the area around it where people who work and shop in the city live is called a (natural region, <u>functional region</u>).

3. Three of the main landforms are plains, plateaus, and (rivers, <u>mountains</u>).

4. Physical features of the land can be used to divide the world into (<u>regions</u>, vertical zones).

5. A region which is based on the kind of government a country has is called a (physical region, <u>political region</u>).

6. Culture regions are based on (one factor, <u>many factors</u>).

7. Countries that do not take sides with the Communist World or the Free World are part of a region called the (<u>Third World</u>, Other World).

8. Countries which use a great deal of technology are called (<u>developed</u>, developing) countries.

9. A country in which people own and operate most of the businesses has an economic system called (socialism, <u>free enterprise</u>).

10. A uniform region is one which has (many features that set it apart, <u>one feature that sets it apart</u>).

B   **Write T if the statement is true. Write F if the statement is false.**

___T___ 1. Geographers divide the world into regions to make the study of geography easier.

___T___ 2. An example of a functional region is Metropolitan Statistical Area.

___F___ 3. The features of the earth's surface are called continents.

___F___ 4. The United States is an example of a physical region.

___T___ 5. Culture includes things such as language, religion, and type of economic system.

___F___ 6. A country in which there were few machines would be an example of a developed country.

___F___ 7. An economic system in which the government owns most businesses is called free enterprise.

___F___ 8. The Cotton Belt of the United States is an example of a culture region.

C   Look at Maps 5–9, 5–10, and 5–11. Write the kind of region shown by each map.

1.   <u>   physical region   </u>

Map 5-9

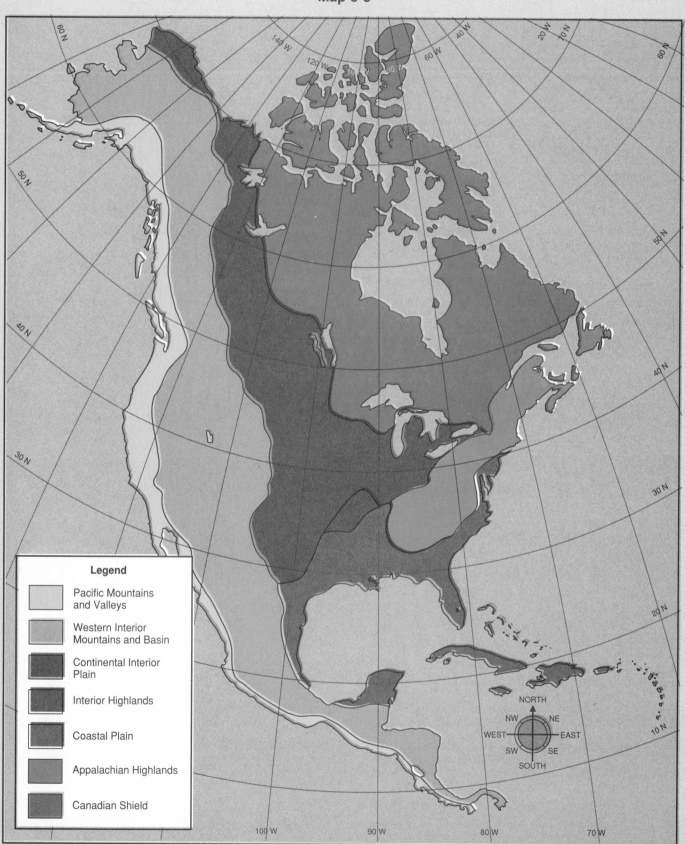

Legend

Pacific Mountains and Valleys

Western Interior Mountains and Basin

Continental Interior Plain

Interior Highlands

Coastal Plain

Appalachian Highlands

Canadian Shield

2. _____ political region _____

Map 5-10

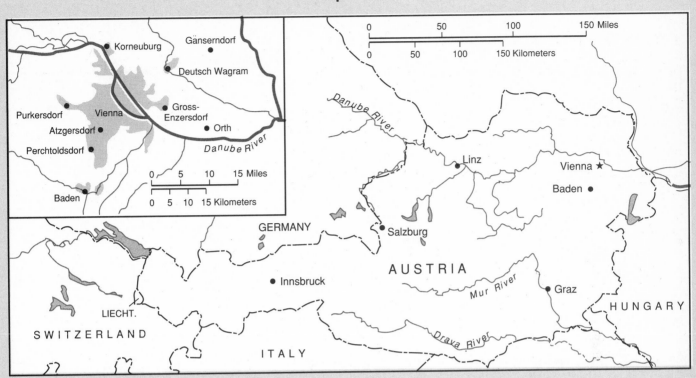

3. _____ culture region _____

Map 5-11

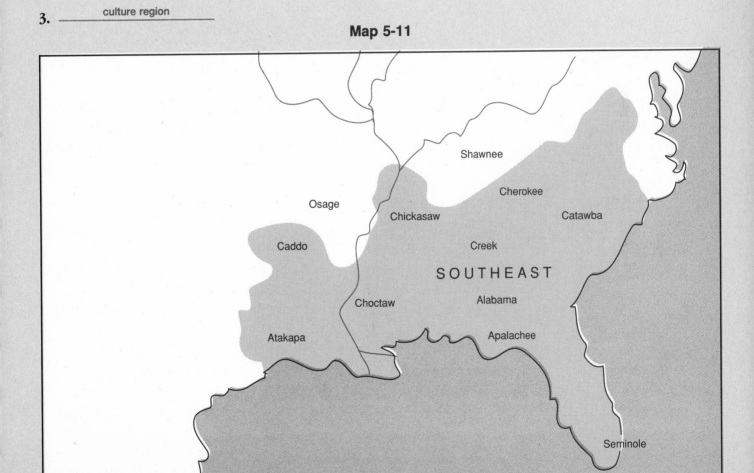

FINAL REVIEW

A   **Use the map on the next page to answer these questions.**

1. Match each country with its capital city.

_c_   Sweden   a. Paris

_e_   Yugoslavia (Serbia &   b. Bern
      Montenegro)
                                                          c. Stockholm
_g_   Spain
                                                          d. London
_a_   France
                                                          e. Belgrade
_d_   United Kingdom
                                                          f. Bucharest
_f_   Romania
                                                          g. Madrid
_b_   Switzerland

2. Match each city below with its position in latitude and longitude.

_d_   Belgrade, Yugoslavia   a. 60°N, 10°E

_a_   Oslo, Norway   b. 64°N, 22°W

_e_   Prague, Czech Republic   c. 56°N, 37°E

_c_   Moscow, Russia   d. 45°N, 20°E

_b_   Reykjavik, Iceland   e. 50°N, 15°E

3. About how far is it in miles between each pair of cities listed below?

_a_   Lisbon, Portugal, to Madrid, Spain
         a. 300 miles   b. 600 miles

_b_   Bern, Switzerland, to Belgrade, Yugoslavia (Serbia & Montenegro)
         a. 400 miles   b. 600 miles

_a_   Bonn, Germany, to Moscow, Russia
         a. 1,300 miles   b. 1,500 miles

4. In what direction would you travel on each trip below?

_a_   from Sofia, Bulgaria, to Helsinki, Finland
         a. north   b. south   c. east   d. west
_d_   from Vienna, Austria, to Paris, France
         a. north   b. south   c. east   d. west
_c_   from Warsaw, Poland, to Bern, Switzerland
         a. northeast   b. southeast   c. southwest   d. northwest
_b_   from Oslo, Norway, to Warsaw, Poland
         a. northeast   b. southeast   c. southwest   d. northwest

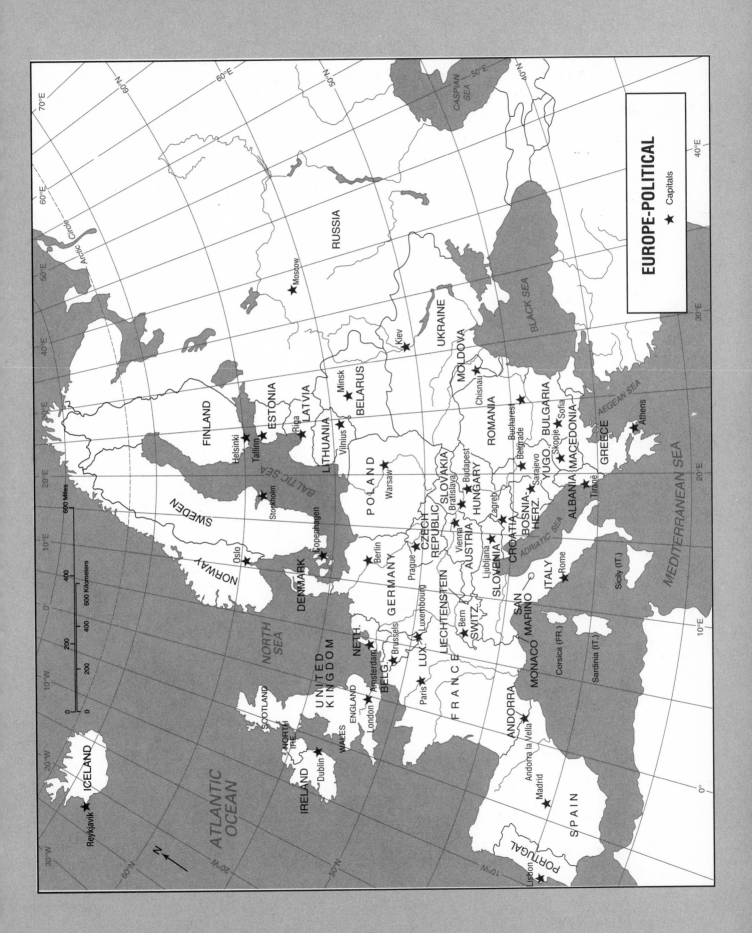

EUROPE-POLITICAL

★ Capitals

B   **Use the bar graph to answer these questions.**

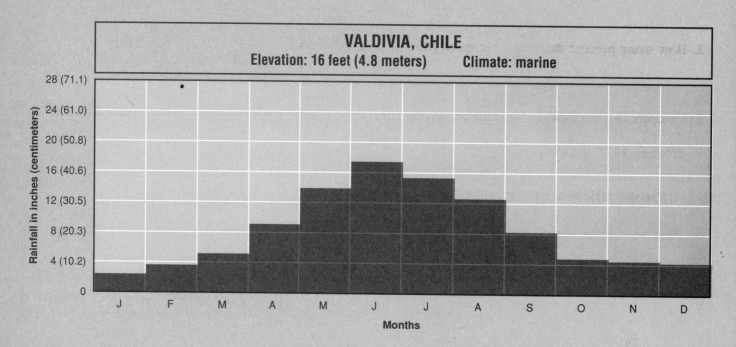

VALDIVIA, CHILE
Elevation: 16 feet (4.8 meters)   Climate: marine

Rainfall in inches (centimeters)

28 (71.1)
24 (61.0)
20 (50.8)
16 (40.6)
12 (30.5)
8 (20.3)
4 (10.2)
0

J F M A M J J A S O N D

Months

_____b_____ **1.** In what month does the most rain fall in Valdivia?
     a. July   b. June   c. January

_____c_____ **2.** In what month does the least rain fall in Valdivia?
     a. December   b. June   c. January

_____b_____ **3.** About how many inches of rain fall in Valdivia in May?
     a. 5   b. 14   c. 9

_____a_____ **4.** About how many more inches of rain fall in Valdivia in August than fall in November?
     a. 8   b. 10   c. 12

C   **Use the line graph to answer these questions.**

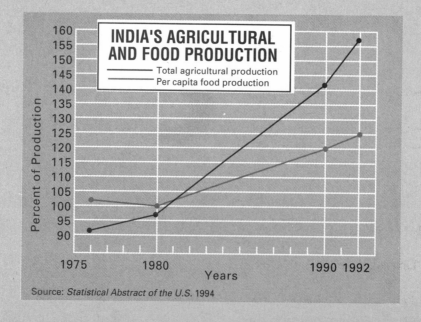

INDIA'S AGRICULTURAL AND FOOD PRODUCTION

Total agricultural production
Per capita food production

Percent of Production

160
155
150
145
140
135
130
125
120
115
110
105
100
95
90

1975   1980   1990 1992

Years

Source: *Statistical Abstract of the U.S.* 1994

1. Did per capita food production in India go up or down between 1975 and 1977?

down

2. How many percent did total agricultural production increase in India between 1971 and 1982?

59 percent

3. Compare the trend of total agricultural production to the trend of per capita food production in India between 1977 and 1982. The trend of total agricultural production was up. The trend of per capita food production was up.

D Use the table to answer these questions.

OIL PRODUCTION IN SELECTED COUNTRIES OF NORTH AFRICA AND SOUTHWEST ASIA

Country	1985 Oil Reserves (in barrels)	1985 Oil Production (in barrels per day)	Years Reserves Will Last*
Algeria	8,820,000,000	1,010,000	37
Egypt	3,850,000,000	910,000	10
Iran	47,670,000,000	2,250.000	58
Iraq	44,110,000,000	1,430,000	84
Kuwait	89,700,000,000	1,070,000	271
Libya	21,300,000,000	1,100,000	53
Saudi Arabia	168,800,000,000	3,730,000	128
United Arab Emirates	32,890,000,000	1,350,000	127

*based on 1985 production levels

Source: *The Middle East and North Africa*, 1987

1. How many barrels of oil did Kuwait produce per day in 1985? 1,070,000

2. How much were Libya's oil reserves in 1985? 21,300,000,000 barrels

3. In what year will Egypt's oil run out, based on this table? 1995

E   **Use the map to answer these questions.**

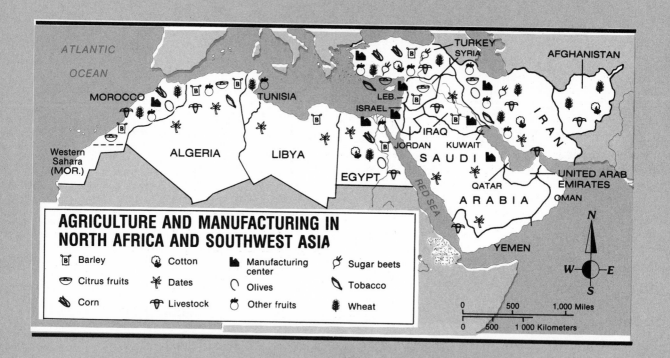

1. What crops are grown in Libya? dates and barley

2. What crops are grown in Egypt? dates, olives, cotton, corn, wheat, other fruits, barley

3. Does Egypt produce livestock? yes

4. What countries produce citrus fruits? Morocco, Algeria, Iran, Syria, Turkey

Glossary

This Glossary contains all of the vocabulary words you have learned in the lessons of this book. The Pronunciation Key will show you the symbols used in the Glossary. An accented syllable is shown with capital letters.

Pronunciation Key

Sound	As in	Symbol	Example
ă	hat, map	a	back (bak)
ā	age, face	ay	Asia (AY·zhuh)
â	care, their	ehr	bareback (BEHR·bak)
ä, ŏ	father, hot	ah	rock (rahk)
ar	far	ar	car (kar)
ch	child, much	ch	China (CHY·nuh)
ě	let, best	e	essay (ES·ay)
ē	beat, see, city	ee	marine (muh·REEN)
er	term, stir, purr	er	pearl (perl)
ĭ	it, hymn	i	system (SIS·tuhm)
ī	ice, five	y	Ohio (oh·HY·oh)
		eye	iris (EYE·ris)
k	coat, look, chorus	k	corn (korn)
ō	open, coat, grow	oh	rainbow (RAYN·boh)
ô	order	or	orchid (OR·kid)
ȯ	flaw, all	aw	ball (bawl)
oi	voice	oy	coinage (KOY·nij)
ou	out	ow	fountain (FOWN·tuhn)
s	say, rice	s	spice (spys)
sh	she, attention	sh	motion (MOH·shuhn)
ŭ	cup, flood	uh	study (STUHD·ee)
u̇	put, wood, could	u	full (ful)
ü	rule, move, you	oo	cute (kyoot)
zh	pleasure	zh	Asia (AY·zhuh)
ə	about, taken	uh	fiddle (FID·uhl)

A

acid rain (AS·id rayn)—rain or snow that carries pollution

altitude (AL·tuh·tood)—elevation

axis (AX·is)—imaginary line drawn from the North Pole to the South Pole

B

balance of trade (BAL·uhns uhv trayd)—the difference in value between a country's imports and exports

bar graph (bar graf)—graph which uses bars to show numbers

birth rate (berth rayt)—how many people are born each year per 1,000 population

C

cell (sel)—area where a row and a column meet

circle graph (SER·kuhl graf)—graph shaped like a circle

climagraph (CLYM·uh·graf)—graph which shows both temperature and precipitation

Communist World (KAHM·yoo·nist werld)—region made up of countries with a Communist form of government

compass rose (KAHM-puhs rohz)—symbol used on a map to show directions

conic projection (KAHN·ik proh·JEK·shuhn)—map projection used for showing small areas midway between the equator and the poles

continent (kahn·tuh·nuhnt)—the largest landmass

contour interval (KAHN·toor IN·ter·vuhl)—amount of elevation between contour lines

contour line (KAHN·toor lyn)—line on a map which connects points of equal elevation

contour map (KAHN·toor map)—map which uses contour lines to show elevation

culture (KUHL·cher)—the way of life of a people

culture region (KUHL·cher REE·juhn)—region based on many factors of how people live

D

death rate (deth rayt)—how many people die each year per 1,000 population

deepwell injection (DEEP·wel in·JEK·shuhn)—the pumping of harmful wastes deep into the ground

degree (duh·GREE)—unit of measurement of latitude and longitude

democracy (duh·MAHK·ruh·see)—a government in which laws are made by leaders elected by all the people

developed nation (di·VEL·upt NAY·shuhn)—country which uses a great deal of technology and has a high per capita GNP

developing nation (duh·VEL·uh·ping NAY·shuhn)—countries in which most of the people still depend on agriculture for their living, and has a low per capita GNP

E

earthquake (ERTH·kwayk)—strong shaking of the earth

economic system (ek·uh·dNAHM·ik SIS·tuhm)—the way a people produce, get, and use goods and services

elevation (el·uh·VAY·shuhn)—height above sea level

elevaton map (el·uh·VAY·shuhn map)—map which shows elevation

energy (IN·er·gee)—the power to do work

environment (in·VYRN·muhnt)—one's surroundings

equator (ee·KWAY·ter)—line of latitude which divides the Northern Hemisphere from the Southern Hemisphere

evaporation (ee·vap·oh·RAY·shuhn)—process by which water is changed into a gas

exports (EK·sports)—products sold to foreign countries

F

free enterprise (free IN·ter·pryz)—an economic system in which people are free to decide what kind of work they will do, and to own businesses and keep the profits

Free World (free werld)—region made up of countries with a democratic form of government

functional region (FUNK·shuhn·uhl REE·juhn)—a large city and the area around it where people who work and shop in the city live

G

Gall-Peters projection (gawl PEE·ters proh·JEK·shuhn) —map projection which shows the sizes of landmasses correctly, but distorts shape

greenhouse effect (GREEN·hows ee·FEKT)—a slow warming of the earth caused by heat being trapped by gases from burning fuels

grid (grid)—set of lines used to find locations on a map

groundwater (GROWND·wah·ter)—water which sinks into the ground

H

hazardous (HAZ·erd·uhs)—dangerous

hinterland (HINT·er·land)—area where raw materials are grown or gathered

hurricane (HER·uh·kayn)—large storms having high winds, heavy rains, and a storm surge

hurricane warning (HER·uh·kayn WAR·ning)—notice that a hurricane is expected to strike a particular location within 24 hours

hurricane watch (HER·uh·kayn wach)—notice that a hurricane is within 24 hours of striking somewhere

I

imports (IM·ports)—products purchased from a foreign country

index (IN·dexs)—alphabetical list of names

interdependence (in·ter·dee·PEN·duhns)— dependence upon one another

K

key (kee)—part of a map which tells the meaning of symbols

L

landfill (LAND·fil)—giant hole in the ground where trash is buried

landform (LAND·form)—feature of the earth's surface

latitude (LAT·uh·tood)—parallel lines on a map or globe running east and west, used to measure distance north or south of the Equator

legend (LEJ·uhnd—part of a map which tells the meaning of symbols

life expectancy (lyf ik·SPEK·tuhn·see)—how long the average person will live

life expectancy map (lyf ik·SPEK·tuhn·see map)—map which shows the life expectancy of people in an entire region

lightning (LYT·ning)—electricity passing between a cloud and the ground

line graph (lyn graf)—graph which uses lines on a grid to show numbers

longitude (LAWNJ·uh·tood)—lines on a map or globe running north and south, used to measure distance east or west of the Prime Meridian

M

manufactured good (man·yoo·FAK·cherd gud) —thing made from raw materials

manufacturing (man·yoo·FAK·cher·ing)—the making of products

map projection (map proh·JEK·shuhn)—a way of showing the earth on a piece of paper

Mercator projection (murh·KAYT·er proh·JEK·shuhn)— map projection which shows true directions and land shapes but exaggerates sizes of landmasses

N

natural resource (NACH·er·uhl REE·sohrs)—something that is found on or in earth

north arrow (north EHR·oh)—symbol used on a map to show directions

Northern Hemisphere (NOR·thurn HIM·uhs·fehr)—part of the earth north of the equator

P

physical map (FIZ·i·kuhl map)—map which shows how the land looks

pie graph (py graf)—graph shaped like a pie or circle

political map (puh·LIT·uh·kuhl map)—map which shows how humans have divided the surface of the earth

political region (puh·LIT·uh·kuhl REE·juhn)—an area that has a particular kind of government

pollutant (puh·LOO·tuhnt)—harmful substance found in the environment

pollution (puh·LOO·shuhn)—something unclean in the environment

population bulge (pahp·yuh·LAY·shuhn buhlj)—people who make up a large group in the population

population density map (pahp·yuh·LAY·shuhn DEN·suh·tee map)—map which shows where on the earth's surface large numbers of people live

population pyramid (pahp·yuh·LAY·shuhn PIR·uh·mid)—graph which shows how the population is divided by sex and age

precipitation (pree·sip·uh·TAY·shuhn)—rainfall or other moisture

prime meridian (prym muh·RID·ee·uhn)—starting point for measuring longitude

profit (PRAH·fit)—the money left over after all expenses are paid

R

rate of change (rayt uhv chaynj)—speed with which change takes place

raw material (raw muh·TIR·ee·uhl)—thing from which products can be made

recycling (ree·SY·kling)—reusing things instead of throwing them away

region (REE·juhn)—a part of the world that is the same in some way

relative location (REL·uh·tiv loh·KAY·shuhn)—location compared to the location of other places on earth

relief (ruh·LEEF)—difference in elevation between places

relief map (ruh·LEEF map)—map which shows the height of land above sea level

resource (REE·sohrs)—something people use

resource map (REE·sohrs map)—map which shows the things found or produced in an area

S

scale (skayl)—relationship between map distances and real distance on the earth

sea level (see LEV·uhl)—average height of water in the world's oceans

self-sufficient (self·suh·FISH·uhnt)—able to meet all of one's own needs

socialism (SOH·shuhl·iz·uhm)—economic system in which the government decides what kind of work people will do, and owns most of the businesses

Southern Hemisphere (SUH·thern HIM·uhs·fehr)—part of the earth south of the equator

special-purpose map (speh·shuhl PER·puhs map)—maps which give one particular kind of information

surface runoff (SER·fis RUN·awf)—water which flows into rivers and oceans

symbol (SIM·buhl)—drawing used on a map

T

table (TAY·buhl)—information presented in list form

technology (tek·NAHL·uh·gee)—the use of tools and skills to make life easier

temperature (TIM·per·uh·choor)—how warm or cool it is

Third World (therd werld)—region made up of countries that do not take sides with either the Communist World or the Free World

time zone (tym zohn)—division of the earth for the purpose of keeping time

tornado (tor·NAY·doh)—most violent storm in nature

toxic waste (TAHX·ik wayst)—things people throw away which can be harmful

trade (trayd)—the buying and selling of goods

transpiration (tranz·puh·RAY·shuhn)—the putting of water vapor back into the air by plants

transportation (tranz·per·TAY·shuhn)—the moving of people and goods from place to place

trend (trind)—how an amount shown on a line graph changes over time

triangular trade (TRY·ang·yoo·ler trayd)—trade between New England, England, the West Indies, and Africa in the 1700s

U

uniform region (YOO·nuh·form REE·juhn)—an area that has one feature that sets it apart

V

vertical zonation (VER·tih·kuhl zohn·AY·shuhn)—change in climate due to altitude

W

waste reduction (wayst ree·DUHK·shuhn)—the making of fewer wastes

water cycle (WAH·ter SY·kuhl)—process by which the earth's water moves from the oceans to the air to the land and back to the oceans

Index

Maps

NUMBER	TITLE	PAGE

Graphs